To Julia ... ...

love

from

Rose Sturgis

THE MAMMALS THAT MOVED MANKIND

A History of Beasts of Burden

R.C.Sturgis

authorHOUSE

AuthorHouse™ UK
1663 Liberty Drive
Bloomington, IN 47403 USA
www.authorhouse.co.uk
Phone: 0800.197.4150

Published by AuthorHouse 09/04/2015

ISBN: 978-1-5049-3945-4 (sc)
ISBN: 978-1-5049-3944-7 (hc)
ISBN: 978-1-5049-3946-1 (e)

DEDICATION

This book is dedicated to my marvellous husband John.

CONTENTS

INTRODUCTION

THE CHOSEN FEW

Elephants and wolves, yaks and reindeer, camels and zebu – what do these motley selections of animals have in common? The answer is that out of the few creatures that Man domesticated six very particular ones were put to work for him – the ones that made mobility easier. Domesticating animals for food and skins was one thing but having them work for us took them and us into new territory. We moved on: we moved on in so many ways. We became no longer dependant on Shanks' Pony.

Mankind, after a shaky start and against all odds, progressed down the millennia increasing in numbers. Certain momentous events affected the human population, which expanded sometimes slowly, sometimes in sudden spurts and surges – never continuously. About 50,000 years ago the extraordinary migration from Africa, and the first use of fire and tools became the chief great events in human development. Then approximately 12,000 years ago, the second significant surge that affected everything came with cultivation of crops and the domestication of animals – the Neolithic Revolution.

The awareness of pre-history is only two centuries old, so our perception of the places animals came to occupy in our lives is only now being appreciated. Before Darwin published "The Origin of Species" all dates in the Christian world were based on the bible and that time scale is very short. Bishop Ussher of Armagh (1581-1656) was convinced that God started making the world in 4004BC on October 23rd, whereas modern geologists calculate the earth is 2-3 billion years old – quite a time difference. Geology actually began with the Greeks but really got going in the C18[th]. Then glimpses of the origins of these animal domestications began to emerge tantalizingly from the mists of pre-history – from fossil remains and archaeological findings, from cave paintings, myths, and early written records. It was science that literally began to open up the secrets of the ancient world and our partnerships with the animals in it.

Contributors to the search and organising of the various data were many. For example Carl Linnaeus (1707-1778) the botanist, physician and geologist and known as the Pliny of the North wrote "Systema Naturae" in 1735. He classified 4,400 species of animals including domestic animals. His best students, known as his 'apostles', travelled all over the world carrying out research for Linnaeus who never left his native Sweden. One of his great contributions was to replace the multi-worded and chaotic taxonomy or categorization which had started with Aristotle with the now familiar binomial, such as *Rangifer tarandus,* the name for a reindeer or *Camilus dromedarius* the binomial for the dromedary camel.

Science is coming up all the time with new techniques to probe into our past, such as radio carbon dating, genetics and DNA, all of which are throwing ever greater light on all our origins. One of the most important – the Radio Carbon Revolution was led by the American Willard Libby in 1947 working at the University of Chicago making it possible to construct the chronology for prehistory by using the measuring of the breakdown of carbon-14 for age determination. This discovery revolutionised archaeology. It became possible to date all ancient civilisations quite independently of each other. For this vital discovery he was awarded the Nobel Prize in 1960. The other equally important discovery was DNA, the hereditary material stored on a double helix in all living creatures. This was discovered in one of the most important breakthroughs by Francis Crick (1916-2004), the molecular biologist, biophysicist and neuroscientist, together with James Watson. Crick was also awarded the Nobel Prize in 1962.

One of the newest techniques, archaeogenetics comes from a study of DNA from archaeological remains plus science based techniques making even more exact chronology possible. This has filled in so much data about the development of all animals including our special six.

Archaeology is the bridge between geology and history.

About 10,000 years ago the revelations began…Man began to see that there was something more to animals than just providing food and skins and so he began to domesticate a chosen few. And so one by one they entered our lives, almost biblically bearing gifts. It was the Old World – Eurasia that had the best selection of animals that could and would be domesticated. Most of these vital domestications took place between 8,500 and 7,000BC, with the exception of wolves/dogs who joined the human pack much earlier.

Out of roughly 4,000 mammals we have domesticated less than a dozen; there was no question of one size fits all. The first rule is self-evident, it rules out all small or weak animals. Then, as far as temperament is concerned, the large cat family is an obvious non-starter; apart from any other regard you don't want to run the risk of your mount eating you! Remember what happened to Miss Riga, in the limerick, who went for a ride on a Tiger! Obviously other dangerous animals like rhinos, hippos, buffaloes and bears are non-starters. Gazelles, the most numerous of animals in Eurasia are prone to panic and are therefore uncontrollable, with the important exception of the reindeer. The diet of an anteater would present a problem. The Ancient Egyptians tried ibex, gazelles and even hyenas. Amerindians tried moose and raccoons but really had nothing to domesticate except eventually the dog. Maybe the Australian Aborigines tried with kangaroos! It was probable they killed all the large marsupials 40,000 years ago – a great pity as it might have been possible to harness such an animal as the *diprotodon* (a ten foot long sort of wombat version of a rhino).

What do we mean by a domestic animal? Nature had bred animals purely for survival. Evolution dished up the raw materials and then we, with our tampering, changed this selection of wild animals into very different creatures. It had become Man's turn to practice his own mini evolutions. He was the new selector; to quote Charles Darwin "The Key is man's power of accumulative selection; nature gives successive variations; man adds them up in certain directions useful to him." We bred them for meat and milk, for speed, size or strength. We bred them for temperament or for looks. Sometimes we aimed for beauty and produced creatures like a thoroughbred horse or a saluki dog; but we also could produce aberrations like a bulldog that can't always breed naturally or a pug that can hardly breathe.

But it must be remembered that domestication was a process not an event. Sometimes it might have been a two-way arrangement with certain cunning animals actually choosing us as likely providers, attaching themselves to human settlements as a form of protection and a means of getting food. Possibly they used submission and the charm of juveniles to attract the humans' nurturing instincts and perhaps in Man awoke the realisation that having these animals in the family could bestow other benefits. But all wild animals have weapons or protection from predators of various kinds so with each there was always a challenge to get them onside.

Hunting was always of the utmost importance but for transport, farm work, and as pack animals a different range of creatures began to be required. Some of those kept for food were suitable for transport, and some of the pack animals could double up as food when they grew old and unable to work. They earned their keep in more than one way. We also began to need animals to help carry our heavier goods and also to carry us for longer distances, and then once we started farming, (in the Neolithic Revolution 9-10,000BC), we needed animals to work in the new fields we had begun to cultivate. So now our association with various cloven hoofed animals began to move onto a different footing. These alliances grew in importance down the centuries. Before riding an animal or pressing him into service as a pack animal or for draught, four main points at least must be considered – strength, temperament, design and availability. Francis Galton (1822-1911), a cousin of Darwin and one of the last of the polymaths, actually worked out six conditions for the domestication of an animal.

1. Must be hardy and be able to bear being separated from its mother.
2. Inborn liking for man and ability to accept him as leader.
3. Comfort loving so happy in captive conditions.
4. Useful to savages as a movable larder, provider of milk, skin etc.
5. Breeds freely.
6. Easy to tend. ('easy to feed' could be included)

He also said that most animals were "destined to perpetual wildness".

A tremendous surge of domestications took place from about 8,500BC to 7,000BC when Man was taking to the pastoral life – not that he had stopped hunting, but a four-footed larder provided a great fall back when game was scarce: the best way to preserve meat is on the hoof. These herd animals were sheep, goats and pigs domesticated in the Middle East mainly to provide milk, meat and wool

However the animals this book is most concerned with are the ones we put to work for us and that narrowed the field considerably. Where did all this start? It started in Mesopotamia, and you may wonder why the ancient Sumerian civilisation took off in such a big way in almost desert conditions. The reason is that until 5,000 years ago Western Asia actually consisted of enormous fertile plains. In 8,300BC the very first crops began to be grown there. But unfortunately it was the very goats and sheep, being kept by man, that by uncontrolled overgrazing slowly turned the 'Fertile Crescent' into desert – an early example of the man's often repeated habit of damaging the environment.

One of the first signs of interference by man to an animal is a variation of colour, white spots, all white, even all black. Difference in hair is another sign; the original horse's mane was short and stiff like a zebra or the Przewalski's Horse; sheep's wool has been made to grow thicker and softer and the outer hairs to disappear. Most animals have been bred smaller to enable their new owners to have more control. The Horse is the one exception as he was bred larger and stronger to use for riding and draught. With certain animals such as dogs, there was a shortening of the snout and jaws and a juvenile look deliberately inbred – neoteny, or as the dictionary defines it 'the retention of juvenile features in the adult'. Part of the price dogs paid for this selective breeding was lower brainpower and volume.

There is a difference between 'domesticated' and 'domestic'. Domesticated are individuals made more tractable or tame but *not* deliberately selectively bred, for example elephants, camels, yaks, water buffalo and reindeer. Camels, yaks and reindeer have this in common – that they are not creatures of agricultural based urban civilisation but are all adapted to harsh physical conditions and a nomadic way of life. But 'Domestic' applies to animals deliberately selected and bred for definite characteristics so that they eventually differ markedly from their original ancestors – for example dogs, cattle and horses; camels have recently joined this list. A species cannot be said to be truly domesticated until it breeds reliably in captivity. 'Feral' of course means domestic animals that have returned to the wild and in some cases reverted in quite a degree to the ancestral stock. But some domesticated animals have been changed to such an extent by human selection that they find it difficult or even impossible to adapt back to wild life.

So in just a few thousand years Man had managed to find an amazing assortment of animals to work for him whose enormous variety had been dictated by geography, altitude, temperature, temperament, migration and most of all – by chance. These six mammals opened up the world for him: now he was able to move across the continents for all sorts of reasons which changed and developed down the centuries – hunting, finding pasture, carrying goods, trade, spreading religions and of course war.

Chapter 1

TURNING WOLVES INTO DOGS

Molussus

"The dog is a gentleman; I hope to go to his heaven not man's."

Mark Twain

What is a dog? Where did it come from? Why is it so important to us and has been since the dawn of time? Our first clue is a study of the Wolf. Man may have used Wolf but Wolf also used Man, and in fact historians are rather at a loss as to whether Man domesticated Dog or whether dog domesticated itself. Whichever it was, cooperation between us and them took hunting, the essential pursuit, onto a different footing.

When did the hunter part of 'Hunter-gatherer start? It is impossible to know but our closest cousins, the Chimpanzees certainly eat meat when they can get it. The gathering part is all too time-consuming as you only have to confirm by watching cows and sheep grazing all day long. So getting meat was very worthwhile, but we needed organisation and cooperation and intelligence to make up for our lack of size and strength. Vitally, that extra nutrition gave us more leisure for innovation and that in turn honed our hunting skills. But to achieve this we urgently needed help with our hunting, and that help was to come.

A day's hunting was one thing but game has a habit of moving on, whether to avoid resident predators or to follow the seasonal food growth, which in turn follows the rain fall. So we were forced to follow the moving larder. But also in pursuit of game were packs of wolves, so gradually a form of co-operation, a kind of symbiosis grew up. That means that our very first mammal friends were wolves who initially collaborated in the hunting and then moved into our lives and gradually took on so many roles and are still doing so.

Every wild animal has some form of defence against predators. Wolves are more dangerous as a pack than as individuals. But once we could convince them we were leaders of the pack then the first big step to domestication was taken. Our first mammal was enrolled. Now we could dig pits and they could surround or drive an edible creature to his death; they could scent the prey and we could help deal with it; and as an extra they could also give warning of predators approaching. We were both after the same thing – a delicious chunk of Mammoth or a haunch of Reindeer. We naturally tended towards the more amenable of the wolves and so gradually bred from them.

To go back to the ancestry of wolves: wolves share a common ancestry with cats, bears, foxes and even hyenas, and they are all descended from Miacis, a sort of weasel like animal that existed 60 million years ago. Then 8 million years ago there were at least two canids in North America, the one called Epicyon meaning 'near wolf' which was more powerful and bigger than a wolf but became extinct. The other in the late Miocene era (10 – 5 Million years ago) was Eucyon, which is Greek for 'original dog' which eventually evolved into the Grey Wolf. Much later, that is 2 million years ago (Oligocene period), the ancestors of the African Hunting or Painted Dog, and the South American Bush dog and also the Indian Dhole split off from the main branch, so that means they are not really closely related to domestic dogs at all although they do superficially resemble them. By the Pleistocene period (half a million years ago) the family Canidae emerged and the more recognisable canine forms of wolves, jackals, coyotes and foxes developed. One of the defence measures of wild animals is a flight distance which they keep between themselves and any threatening creature. This distance would have become shorter and shorter as Man and Wolf began to find cooperation a rewarding reaction to each other.

Charles Darwin thought that dogs were descended from a mixture of wolves, jackals and coyotes. Konrad Lorenz, the author of "King Solomon's Ring" (1952) also thought much the same. There are approximately thirty-six wild canines so that produces quite a wide choice of ancestors. However in 1997 an international team of scientists led by Robert Wayne at the University of California compared the mitochondria (the female line of DNA) of wolves, coyotes and jackals. They took dozens of samples from many different breeds of dogs and compared them with samples from wolves and coyotes and found that the mitochondrial DNA of wolves and dogs differed by only 1%, whereas wolves and coyotes differ by 6%, strangely enough, about the same as do some different races of Man. This meant that the wolf and *only* the wolf is the ancestor of the dog. Lorenz would have been fascinated by the new technique and its findings but unfortunately he died in 1989. There seems to have been four species of wolves in the canine ancestry, of which the Grey Wolf was once the most widely distributed mammal, showing their success. Just recently Olaf Thalmann of the Finnish University of Finland has confirmed the four genetic lineages are all North European.

When did this domestication take place? When did this partnership begin? Archaeology seems to place humans and wolves together from 100,000 years ago or even earlier and wolf-dogs from about 30,000 years. This has recently been confirmed by Finnish and German research. Wolves being pack animals care for each other's young, so all it needed at the beginning was for the wolf/dogs to accept a man as an Alpha Male.

The earliest signs of dogs are Rock Art and skeletal remains that indicate that by 14,000 years ago Dogs ranged from North Africa right across Eurasia and even into North America. They were certainly being used by Mesolithic, maybe even Palaeolithic mammoth hunters of Ukraine. The earliest remains of a dog's skull was found in the Kesserloch Cave in Switzerland and dates from the Magdalenian period about 14,000 years ago. Bones of Dogs and Wolves have been found at Maglemosian sites in NW Europe dating from the Mesolithic or Stone Age (6,000-1,500BC). Other early remains of a domestic type dog have been found in the Middle East, where a grave with a dog's skeleton was found in Northern Israel in a settlement called Ein Mallaha, dating from 9,000BC and others have been found in Iraq and Jericho dating from about 6,750BC. Indeed, by the time man reached Denmark in Europe in Neolithic times he certainly had domesticated dogs accompanying him; this is shown by dog remains having been found there, dating from 8,000BC. Further proof showed up in 1976, when the Hebrew University Centre for Prehistoric Research excavated a campsite dating from 10,000 years ago and came across the skeletons of a child clutching a five month old puppy.

And so *Canis familiaris* evolved – our very first animal partner. To this day dogs and wolves can easily interbreed producing wolf-dogs that show typical hybrid vigour. The choice of dog mate would obviously have to be a close relation and big such as a Malamute or a Husky, not a Chihuahua or a Pekingese.

How did these animals spread across the continents? About 18,000 years ago during the last ice age, the ancestors of the Amerindians and the Inuit crossed over the Bering Land Bridge that joined Asia and North America from Asia and may have taken semi-domesticated dogs with them. Other animals which used the Land Bridge, which measured about 1,000 miles north to south, were wolves. The earliest actual dogs as opposed to wolves found in North America date from 1,500BC. The dogs found in Stone Age human settlements were very similar to today's Eskimo dogs. This shows that dogs have had thousands of generations in which to undergo the genetic changes that have led to such a wide variety of different forms.

There is a perfect encapsulated example of manmade, accelerated evolution in the "Tame Silver Fox Experiment", performed by Dmitri

Belyaev in Russia, in the last century. In less than fifty years – only nine generations – selective breeding resulted in animals which behaved, smelled and sounded far more doglike than fox like. Even their appearance had altered into mottled colours, floppy ears and curly tails and they became totally compliant with humans. Belyaev reckoned that amenable behaviour was the yardstick not size or appearance. Not only their looks changed but their habits: they barked like dogs and had grown to wag their tails when pleased. Even their ears drooped as it was no longer required to prick them up, at the signs of danger. This was a transformation – an evolution in a nutshell. This demonstrates that the magician that transformed wolves into dogs was Man.

Hunting with dogs – the first and most important thing that had brought Man and Wolf together, was slowly advanced by selective breeding but there was a very long way to go from the barefoot men wearing skins and running alongside wolves to men in Pink coats riding horses to hounds blowing horns and stopping for a stirrup cup.

The actual hunting of game with hounds dates back to Assyrian, Babylonian and Ancient Egyptian times and was known as venery. Remains of *Canis familiaris* dating from 4,700BC have been found in Egypt, probably of dogs brought over by Neolithic herdsmen from the Middle East into North Africa. Down the centuries hunting with dogs improved with selective breeding. Man-made evolution was on its way.

An early example of those specially bred hunting dogs was the Ibizan hound probably descended from the 'Tesem' an ancient Egyptian hunting dog whose pictures can be seen on the walls of Ancient Egyptian tombs and pyramids and was used for hunting by the Pharoahs from 3,000BC. It was probably introduced to the Balearic Islands, especially Ibiza (or Eivissa), by the Phoenicians in about C8[th] BC when they were trading round the Mediterranean. Like all sight or gaze hounds it obviously had to have extremely good eyesight: this was helped by its long jaw and neck giving it stereoscopic vision. It also has very big easily recognised pricked-up ears. The other main requirement clearly was speed: this was aided by long legs, a flexible back and a long stride

In the same category belongs the Saluki, the royal dog of Egypt, reckoned to be a breed at least 8,000 to 10,000 years old, which makes it one of the fourteen oldest breeds of dog. Arabian literature states that it originated in the city of Saluk in the Yemen. They appear on Sumerian rock carvings in Mesopotamia dated about 7,000BC and also on Egyptian tombs dating from 2,100BC and were obviously highly regarded as they were often mummified with their owners. One method of hunting was to use them in partnership with falcons who would find the prey and then the Saluki pack would hunt it down. St John Philby (1885-1960) the great English explorer and Arabist and also known as Sheik Abdullah was one of the first men to cross the Empty Quarter. He described many encounters with salukis during his Arabian travels and contacts with Berbers and Saudis.

They were usually carried around on their camels and horses until a quarry was sighted. A gift of a saluki was always considered a great prize. The Egyptians had depictions of real Saluki-type hounds on ceramics from 3,500BC and on tombs dating from 1,400BC. Cave drawings of men hunting with dogs were recently found about 25 miles SE of Cairo dating from approximately 7,000BC. They were first imported into England in 1840 but did not really take off as a breed until the Hon. Florence Amherst imported one in 1895 that had been bred by Prince Abdullah of Jordan.

For hunting more dangerous prey, the Greeks and the Romans employed a much more massive dog called a molossus, capable of tackling wild boar; the Assyrians hunted lion with special mastiffs. There are divergent views on whether the molossus was the ancestor of the mastiff or whether it was rather more the greyhound type, slender, fast and long-legged. The Great Dane may even be descended from it as it used to be called the great boarhound.

Another very ancient breed used by the Greeks for running down the gazelle or for coursing the hare is the greyhound or gaze hound, which needed great speed. It is thought that the name is a corruption of 'Greek Hound'. They might even have been introduced into England as early as the C4[th] by Phoenicians, again trading, but this time for Cornish tin. Other very early forms of the Greyhound type were the Irish wolfhound and the Scottish deerhound both found in Britain. The Irish wolfhound dates to at least as far back as the C1[st]AD and was used not only for hunting boar and wolves and deer but to guard the farmhouses and also for war. The Irish motto concerning them was "Gentle when stroked, fierce when provoked!" They were considered very much royal dogs and were frequenters of the halls of the Irish Kings and their striking presence is still used for certain ceremonies by the Irish Guards. The present incumbent is called Donnachadh and like all his thirteen predecessors, is named after the High Kings or legendary chieftains of Ireland.

The Scottish deerhound is also believed by some to have existed in a time before recorded history, possibly bred by the Picts and Scots and is closely related to the Irish wolfhound. There is actually a battered carving of them being used in hunting on an ancient stone dated about 600AD found in Ross in the far north of Scotland called the Hilton of Cadboll Stone, now in the Museum of Scotland. Amongst the hunting field depicted is, rather surprisingly for the times, a woman riding side-saddle. The main quarry in those times was red deer, but the breed fell on hard times when shooting and stalking took the place of hunting and they were no longer required.

New rules and regulations were brought over from Normandy to England by William the Conqueror in 1066. He was passionate about hunting and introduced Forest Laws – the Laws of Venery were to protect all beasts of the chase from the peasants and restricted hunting to the King and the nobility: the punishment for infringing these laws was often death, or at least being blinded. The term 'forest' denoted far more than a heavily wooded area as we interpret it. Over a third of the land, by the C13th was designated Royal Land. William the Conqueror had introduced this concept when he invaded England but less than 200 years later, Magna Carta (1215) limited the King's rights in five famous clauses and by the mid C17th the Laws of Venery had more or less died out. So from the Middle Ages hunting was moving into a new phase from a necessity of life to a sport practiced by the nobility, and a training ground for Man's obsession – War.

But although we now had bred dogs to help us hunt, perversely one of our favourite quarries centuries later was still the dog ancestor – the wolf. It was hunted partly to protect livestock and partly for its pelt, even though it had a foul odour. In the C9th Charlemagne founded an elite corps of wolf hunters called "Luparii". In Saxon times names with the link to wolves were very popular. Examples are Ethelwulf (Noble Wolf), Berthwulf (The Illustrious Wolf), Eadwulf (Prosperous Wolf) Ealdwulf (Old Wolf.) This reflected the brute power that showed off man's greatest distinction and machismo in those days. Then Edward I of England (r. 1272-1307) ordered their extermination. They finally became extinct in England under Henry VII (1485-1509) but lived on in Scotland. Before its extinction, it was considered by the English nobility as one of the five "Royal Beasts of Chase" – Hart, Hind, Boar, Hare and Wolf. How strange it seems, that they continued to wipe out the very animal that they found the most exciting to hunt. In those days animals did not have the same fear of man as they have learned to have now and the flight distance would have been shorter.

In the Galapagos, where the birds have yet to learn the terror of man you can stand or sit right up close to boobies, albatrosses or the iguanas, without them worrying in the least. It was this lack of fear of course that caused the extinction of the Dodo and the Moa and all the other creatures we have exterminated. So in earlier times, when hunting, it was possible to get nearer your prey than we can in present times. At the very least you would think that it would be obvious that killing off your favourite food creature would lead to there being no more of them to eat. But no! We are still doing the same thing today with the North Sea Cod and the Bluefin Tuna.

A report of hunting at the other end of Asia was made by Marco Polo who was attached to the court of Kublai Khan in the C13[th]. There he observed that a very close human–canine bond existed and that the emperor kept 10,000 hunting dogs attended by the same number of men.

Another very well known and beautiful sight hound is the Borzoi, bred by the Russian Czars and aristocracy and very tall – up to 28inches at the shoulder. First mentions of them are from the C13[th] and in 1613 The Imperial Kennels were founded. Borzois were favourite presents to bestow on visiting royalty. Like everything else of the Czarist times, hunting took place on the grandest of scales with at least a hundred Borzois in the pack. Scent hounds were employed to track down the wolf with the hunters on horseback, then when the prey was sighted, the Borzois were unleashed to hold the animal down until the hunters arrived to deliver the 'coup de grâce'.

During the reign of Edward IV (C15[th]) a very early book on Hunting Hawking, Fishing and Heraldry entitled "The Boke of St Albans", written by Juliana Barnes (or Berners), born about 1388 who became the prioress of St Mary of Sopwell, near St Albans, in Hertfordshire. She writes about 'the wulfe, the harte, hare and the boore'. She was a great huntress and considered to be the Diana of her time. Another one of the earliest book on dogs in the English language was entitled "Of Englishe Dogges" written in 1576 by Dr Johannes Caius, who when describing greyhounds wrote "Some are of the greater sorte, some of a lesser; some are smoothe skynned and some curled, the bigger therefore are appointed to hunt the bigger beastes, the buck, the hart, the doe." Several peers, kings and even emperors wrote books such as "La Chasse Royale" by Charles IX of France.

Dogs were now definitely being separately classified as scent or sight hounds. Scent hounds of course had different requirements. For a start, speed was not important but endurance was. Other characteristics are large noses with open nostrils, loose moist lips and large long ears to help scoop up the smells.

A very ancient example of a dog that uses smell to a remarkable effect is the Bloodhound, as it can follow days old cold scent over huge distances. Its ancestry goes back to the 700's and the St Hubert dogs in the Ardennes region of Belgium. These dogs, named after the patron saint of hunters were crossed with the Talbot hounds and came to England in the time of Queen Elizabeth, whose favourite, the Earl of Essex is supposed to have kept 800 of them. The origin of the name comes from the dog's connection to royalty and the best blood lines. Their modern career began in the 1600's when they were used to track sheep stealers and poachers. They are still used today by the police to track down criminals and they have the reputation of being able to pick up a scent that is 100 days old, and follow it relentlessly for 100 miles, so their other name is sleuth hound.

Probably the best known of the dogs that hunted by following the scent is the English Fox hound. Wolves, now extinct in England, still lingered on in great numbers in the wilds of Scotland and were considered the most exciting prey of all to hunt. Hunting had always been a most vital reason for moving overland but after domestication of edible animals the Chase was partly to keep down vermin such as wolves and foxes, partly to get game for eating such as the red deer and wild boar, but also very much for sport. By the late 1500s there was also a shortage of deer caused by over hunting and this led to the perception that a new prey was needed, and so the fox was selected. During the reign of Henry VIII, the English Fox hound was deliberately created by a careful mixing of the Greyhound for speed, the fox terrier for its hunting instinct, and the Bulldog for tenacity in the hunt. Another story of its origins is that it is descended from the heavier hounds bred by St Hubert, the patron saint of hunting, and brought over during the Norman Conquest and then mixed with the Talbot hound, now extinct but remembered in heraldry. Studbooks for Fox hounds were kept as early as the 1800s. They were specifically bred to trail foxes and live around horses for which purposes they are still used. Now there are fox hunts all over the world – France, Italy, USA, Canada, Australia and Russia.

Other variations bred for hunting not only foxes but hares too are Bassets and Beagles who also use scent to hunt down their quarry. The Basset, dating from the 1500's has been deliberately bred with very short legs so that people on foot can keep up with them while hunting rabbits. Their name comes from the French *bas* meaning low. The Beagle was also a deliberately bred smaller version of the Fox hound for much the same reason and apparently Elizabeth I owned Pocket Beagles, which must have been really tiny.

In Spain the Spanish Pointer was the hunting dog and it was the predecessor of all other pointers. As far back as the early C17th the English version was being used by sportsmen to find hares and then for greyhounds to chase them. They seem to be a mixture of greyhound, fox hound, bloodhound and bull terrier. They really excel at flushing birds out of their cover, and are extremely popular in the United States.

Also in the C17th hunting was a great sport in Maryland and red foxes were actually brought over and introduced into the countryside to give them a prey for hunting. Later, in the C18th, George Washington and Thomas Jefferson both had packs. The American ideal was to enjoy the chase and not go for the kill. In Australia the introduction of red foxes for hunting had a devastating effect on local animals, as so many of the introductions of new animals have had in Australia – what a list!...Rabbits, camels, pigs, cane toads, so now shooting foxes, as well as hunting has become part of the control.

Hugo Meynell (1735-1808) is considered the father of modern hunting in England; he was Master of the Quorn Hunt for 47 years and developed a faster breed of hounds, with a better nose for scent and greater stamina.

The hunt was named after his home Quorn Hall in Quorndon, North Leicestershire.

Anthony Trollope insisted that there was social inclusion for everyone from attorneys and bankers to butchers and bakers. Robert Surtees' famous creation Jorrocks the Cockney grocer remarked "'Unting is all that's worth living for – all time is lost wot is not spent in 'unting – it is like the hair we breathe…" Yet there is no doubt that part of the reason for the hunting ban that was to come was inverse snobbery.

Curiously hunting was banned in Germany right back in 1934 by Göring as commanded by Hitler, the vegetarian, whose kindness to animals did not extend to the human race.

In England a ban on Fox hunting was proposed as long ago as 1945, and was actually banned in Scotland in 2002, and after several attempts it was finally got through the English Parliament by the Labour Party, led by Tony Blair and pushed by his wife Cherie after the Burns Inquiry in 2004. This had the unexpected but defiant English reaction result that hunting became even more popular and there are still 184 active hunts, which to stay in business, have to conform to complicated rules. There were found to be 200 packs killing 21,000 / 24,000 foxes a year. Many more are dug out and shot. All together there are estimated to be 217,000 foxes in the United Kingdom.

But Man had realised that hunting was not the only use to which a dog could be put. From early times they had acted as guard dogs then later they had been trained to herd the newly domesticated sheep or goats (8,000BC).

Dogs also had a role to play as beasts of burden and draught animals. In North America, the horse having long gone, dogs were the only beasts of burden. The invading Spaniards in the C16[th] observed large dogs carrying loads of up to 40lbs or more on their backs. They also dragged a device known as the travois. This consisted of two trailing poles eight to ten feet long, made of hardwood in the shape of a V forming an isosceles triangle. This was lashed one on either side of the dog with the other ends of the poles dragging on the ground; burdens were placed on a platform that bridged the two poles and were often made of netting. The wheel had yet to arrive in the Americas but the travois was actually more efficient for dragging loads over boggy or sandy ground than wheels would have been and the dogs were able to drag loads up to 250lbs. Women were in charge of their construction and sometimes when moving camps children were put in willow cages to be drawn by the dogs. Sometimes two dogs lashed apart were used to drag old or sick people.

The Northern tribes, especially the Athabascans of Alaska, kept literally hundreds dogs as beasts of burden. They were used for hunting buffalo and were also pets and companions. Even when horses were introduced, dogs were still employed right up to the C19[th] as they were so much cheaper to acquire and feed.

The travois was one thing but for snow and ice the sleigh was the great innovation. In NE Siberia a tribe called the Chukchi gradually over hundreds of years, bred Siberian Huskies from wolves, incidentally making them another one of the oldest breeds of dogs. They were trained to draw the sleighs that took the men to good areas for fishing, whaling and hunting walrus. Huskies were exceedingly robust and could travel up to a hundred miles in a day. The raising of them became a family affair with the women being in charge of choosing which dogs were for breeding and which were to be neutered. This was not only to select the bloodlines but also because neutered dogs survive on less food and are far less aggressive – the last thing you want when miles away on the ice in a howling cold wind is a dog fight. But the huskies had yet another role and that was to keep the family warm at night by acting as sort of canine duvets in cold that could plummet to -100°C. The temperature of a night would be judged in terms of how many dogs were needed to keep you warm as a 'two dog night' or a 'three dog night'. Life of the Chukchi totally revolved round their dogs, which also acted as companions for their children and were also a centre of their religious life. They believed that two Huskies guard the gates of heaven and that they would turn away anybody that has shown cruelty to a dog in their life time. Other similar histories of breeding are the Malamutes of Alaska and the Samoyeds of Russia. The latter not only pulled sleighs but herded reindeer and helped hunting wild animals including polar bears.

Using dogs for pulling sleds or sleighs still continues and has been developed into races. In Aviemore they celebrated in January 2013 their 30th annual race run by the Siberian Husky Club of Great Britain. It was run near the Cairngorms on forest trails round Lake Morlich and consisted of 1,000 sled dogs and 250 Mushers; a Musher is the name given to the driver. Unfortunately owing to the scarcity of snow they sometimes have to use a form of tricycle sled with three wheels. Apart from Huskies, other dogs used are Malamutes, Samoyeds and even Standard Poodles.

But the greatest sled race is the Iditarod run from Anchorage to Nome on the Bering Sea a distance of about 408 miles. It is known as the Last Great Race on Earth. It all started when William Goosak in 1908 brought teams of Siberian Huskies from Siberia to Alaska. Apart from racing, in 1925 diphtheria serum was rushed from Nenana to Nome by dog sled to save the town from an epidemic. Amundsen got his dogs that took him to the North Pole from Nome.

However further south and below the snowline Man was beginning to use dogs for various kinds of draught work. A more universal form of transport than the travois and the sled was enabled in Europe by the invention of the wheel. Dogcarts began to be used by people who couldn't afford horses or donkeys. They were still in use on the continent until quite recently, with one, two or more dogs delivering bread or milk in Belgium and France. There was the double advantage of dogs being able to guard the contents of the carts. However in England, in an early form of animal rights, dog carts were banned in the early C19[th], and in 1839 'The Dog Cart Nuisance Act' was passed prohibiting the use of dog carts within a radius of 15 miles from Charing Cross. Nothing was mentioned of horses or mules performing the same function. As usual with ill thought out laws based on sentiment and prejudice there was an unintended reaction – many dog owners were too poor to keep a dog that didn't work for its living and so they were turned out into the streets to starve in their thousands.

The Newfoundlands are another breed of dogs capable of draught work. They are related to mastiffs and St Bernards. They come in two sizes the Greater and the Lesser or St John's Water Dog. The former pull carts and being web-footed are also used to take the lines from shore to ship when they were docking, and to pull fishing nets. They have the most amazing propensity to swim out and save sailors from drowning. One famous rescue was by Ann Harvey and her father and brother and their dog Hairy Man who saved 160 Irish immigrants when their ship was wrecked off the Newfoundland coast. In 1820 the Earl of Malmesbury imported some to retrieve ducks at his duck shoot and so began the Labrador breed.

The other early rescue dog and relation of the Newfoundland, the St Bernard, are the heaviest dogs in the world. They were bred from the Tibetan Mastiff, the Great Swiss Mountain Dog and other farmers' herding dogs, and named after the Hospice at the Great St Bernard Pass and first mentioned in the C17[th]. The monks living there were famous for rescuing people lost in the snow, and using dogs to track them down. One legendary dog was Barry or Berry (1800 -1814) who is credited with saving well over forty lives. The other talent of St Bernard dogs was being able to detect the advent of bad weather, very useful in the mountains. Of course they never took the legendary barrels of brandy with them as that would have been the very worst thing for hypothermia.

Having guarded cattle in war, dogs were also employed to guard property and domestic animals in peace, the most popular breeds being not only the Rottweiler but also the Dobermann Pinscher and the German shepherd. Some of these breeds started out as hunting dogs and then moved on to this new work. Black has always been the preferred colour as it is considered to be more intimidating to would be invaders.

The Greeks had developed a few breeds of dogs particularly the Laconian hounds and the heavy Molossians, but it was the Romans who were really in the forefront of dog breeding and used dogs in warfare, one per legion. Then when they arrived with their molossian dogs in Britain, they found an even fiercer, more massive dog *Pugnaces Britanniae* which they quickly incorporated into their breeding programs. Even in the Middle Ages Attila the Hun took Molossian dogs into battle, but now as a breed they seemed to have ceased to exist. There is a magnificent statue of a Bull Mastiff type or maybe a molossus from C1[st]AD in the Uffizi Gallery in Florence.

The enlarging Roman Empire produced more and more wealth for some people and this led to new luxuries and fashions and as a result the reasons for owning a dog also changed. No longer were dogs required purely for utilitarian reasons – to hunt, to guard, to fight, to carry and drag goods. Now there were more light-hearted reasons to own a dog. They appear in Roman mosaics and Assyrian sculpture and already there were an assortment of sizes, colours and shapes owing to the fact that the Roman Empire was spreading everywhere and picking up other people's breeds. This deliberate selective breeding was becoming more and more developed and gave rise to early forms of greyhounds, dachshunds and even lapdogs. So by this time dogs were not only bred to be useful and hard working, some were definitely bred as pets.

In China by the C1st there was even an early Pekingese (or Foo Dogs as they were called) which acted as companions to the Imperial ladies and the palace eunuchs. These demonstrate an early example of paedomorphism or the deliberate creation by selective breeding of a juvenile face with a shorter jaw and more crowded teeth, as can be seen not only in Pekingese but in pugs and bulldogs. Unfortunately it also leads to a decrease in the brain size, which can be by as much as 20% to 30%. Roman ladies had lapdogs literally to sit on their laps for comfort and according to Pliny the Elder (23AD-79AD) who actually mentions a Maltese lapdog in his book *"Naturalis Historia"* it was thought that their warmth could cure a stomach ache. He also warns that madness in dogs is dangerous to human beings when Sirius, the Dog Star is shining, and it is then that it causes hydrophobia (rabies). So it was considered a wise precaution in those days to mix dung, perhaps with that of fowls, in the dog's food, or if the disease has already taken hold, hellebore.

Other new roles for dogs began to emerge in mediaeval days – using them for work; there were small dogs to turn the turnspits in the kitchens, there were larger dogs to deliver meat for the butchers, there were even dogs trained to turn wheels, others to draw water from the well and there was even a Tinker's cur that was trained to carry his tools in a sort of dog back pack.

The difference in size between the tiny Chihuahua and the giant Irish wolf hound is literally down to one gene. The penchant for toy dogs from Roman times was passed on to the royal families of Europe. The Papillion is shown in paintings of the French royal court down the centuries by such masters as Watteau and Fragonard. There is the legend that Marie Antoinette carried one under her arm to the guillotine, but like the Napoleon story it is probably apocryphal. One theory is that its origins go back 700 years, which would make it one of the oldest breeds of dogs. Another favourite was white fluffy Bichon Frise dog which became established in the royal court of Francis I of France (1515-1547). Henry III was said to be so enamoured with these dogs that he fashioned a basket to be placed around his neck in which he carried his pets. They appear in

Paintings by Goya (1746-1828), particularly the famous one of the Duchess of Alba or the White Duchess, whose dress matches the dog's fur. Another dog immortalised by Goya in the C18th and before him by Albrecht Durer (1471-1528) is the poodle. It was originally a German water dog and its very name comes from the High Dutch meaning 'to splash in the water'. However it was standardised by the French and became a very good gun dog. Its intelligence made it popular in circuses and with the gipsies. They hardly shed hair from their curly coats which make them popular in households particularly where there are allergies. And lately they have been crossed with Labradors to make a Labradoodle which with the kind heart of the Labrador and the intelligence of the Poodle make marvellous dogs for helping disabled people as is shown more fully in the last chapter.

A rather moving story comes out of the 2nd Opium War (1860) when the Forbidden City was invaded by Allied troops and the Emperor Xianfeng fled, leaving his elderly aunt and five Pekingese behind. The aunt committed suicide but the lucky dogs ended up in extremely aristocratic situations with the Dukes of Wellington and Richmond and even Queen Victoria. Almost unbelievably the DNA of these flat faced little lion dogs is very close to the wolves' DNA. 'Lion Dog' is an apt term for them for they are not just toys but have great character and courage. One of the few survivors of the sinking of the Titanic was a Pekingese.

Dogs may have originated in the Northern regions of the world but like every animal they migrated and some gradually moved south. They moved round and across the Mediterranean and spread down through the Sahara, which, in those days was not a desert but green savannah. The Early Iron Age Bantu brought them further south and remains of them dating from 570AD have been found as far south as South Africa. *Canis africanus* were the dogs that came down through Africa with Arab traders and are recognised as a 'breed' called 'AfriCanis' which can be found in African households all over Southern Africa.

Another interesting Dog story is that of the Dingo. Their almost identical relatives live in Thailand so they were more than likely brought down from Eastern Asia about 3-4,000 years ago and arrived in Australia in late Palaeolithic times. They resemble the Indian Pariah dog and are usually a golden brown; they have unique wrists on their front paws that can rotate almost like ours, which means they can turn door knobs; strangely they cannot bark but only howl. They are not easy to domesticate but now are interbreeding with feral dogs, and may soon be lost as a 'pure' breed. They have a definite place in the Aboriginal culture, which seems reasonable as it was the ancestors of the Aboriginals who brought over the original dingoes. But the European Australians' views vary from the one extreme of wanting to exterminate every last one to the other of wishing to reintroduce them into areas where they no longer run. The Aborigines have, of course been in Australia for at least 40,000 years and also have a dual attitude to Dingoes. As puppies they were kept as pets and even wet-nursed by the women and given names. They were good watch dogs but in time of food shortage were sometimes eaten, and they are also used as body warmers on cold nights in the same way the Chukchi used their Huskies. However, on maturity they would push off back to the wild and get on with reproducing the next generations. Rather curiously the Aborigines never used them for hunting, probably because the Dingo is not a pack animal; its habit of hunting alone or in a pair is more akin to a jackal, so they do not develop a fealty to their pack leader, the alpha male dog or man, like the wolf did. However the Aborigines used to follow the sounds of Dingoes on the hunt and if they were in time would snaffle the meat; and since the introduction of dogs from Europe they do hunt with cross breeds and go after kangaroos and other marsupials to satisfy their hunger for meat. The Dingo is a great hunter and most likely caused the extinction of the Thylacine, the Tasmanian Tiger which although a marsupial it looked extraordinarily like a dog with tiger stripes. The longest fence in the world was created to protect sheep from the predations of dingoes. The new Zealanders had their own version of the dingo – the kirri. Another version is the New Guinea Singing Dog (*Canis hallstromi*) which was found in Thailand, Sulawesi as well as Papua New Guinea. Originally thought to be a new species they have been demoted to yet another feral dog, but a society for their preservation has been formed in Oregon, North Carolina.

Wolves and dogs can naturally interbreed and the hybrid is known as a 'Lycisca'. Coyotes and dogs also sometimes happen to interbreed and produce "coydogs" (product of a male coyote and a female dog) or "dogotes" (male dog plus female coyote). These hybrids can cause problems as the dog side of them causes a lack of fear of humans coupled with the cunning of the coyotes.

One unexpected consequence of all this breeding is to provide a moving picture of evolution. Darwin used the many breeds of pigeon to illustrate his theories but dogs could just as easily have provided him with an illustration. A strange coincidence is that the first dog show ever to be held in England – in Newcastle was in 1859, the very year that Darwin published "The Origin of Species".

Many breeds had been stabilised in the 1800s and there are now more than four hundred breeds of Dog, classified into roughly six groups: hounds, working dogs, sporting dogs, terriers, toy dogs and unclassifiable pets. Thanks to the plasticity of the canine genome dogs can vary enormously in size and shape. Variety in breeding has gone mad – all sizes and shapes, pricked ears, floppy ears, a huge range in the quality and colour of hair. The tail also comes in many forms, smooth, fluffy and with some breeds it is carried curled over the back. The reason behind the ability to breed such a variety of dogs is because they have so many chromosomes – seventy-eight compared to humans with only 46. This gives the breeder so much more material to deal with.

In England awareness had grown that it was becoming necessary to have some sort of control or legislation about dogs, breeding and other aspects. This led to the foundation of the Kennel Club in 1873 by 12 Victorian gentlemen. At first it was formed to regulate dog shows and produce a stud book. But later it became concerned about dog health and the results of selective breeding. A more recent example was a programme related to hip dysplasia in 1965.

In 1911 the Fédération Cynologique Internationale, a World Canine Organisation with its headquarters in Belgium was created with the aim to promote and protect cynology and purebreds 'by any means it considers necessary'. The word 'cynology' sounds scientific but does not actually appear in dictionaries, and cannot be considered to be a field of science. The federation originally included Germany, Austria, France, Belgium and the Netherlands. It naturally disappeared during WWII but was recreated in 1921 and now claims 87 members and recognises 343 breeds each of which is considered the property of a specific country. Standards have been drawn up on which judges base their verdicts at shows. They

18

also are concerned about in-breeding, cruelty to dogs and hip dysplasia, a hereditary dog disease.

Now so many breeds have been distorted by weird breeding practices that there is quite a debate going on in the dog world, and Crufts is at the centre of it. Crufts was founded by a general manager of a dog biscuit manufacturer called Charles Cruft and he held the first dog show in 1886 for which there were 600 entries. In 1891 the show had moved to the Royal Horticultural Hall in Islington, and had 3,000 entries and prospered under royal patronage. Cruft died in 1938 and a few years later the show was taken over by the Kennel Club, and is now held at the National Exhibition Centre in Birmingham. By 1991 it was recognised by the Guinness Book of Records as the biggest dog show in the world. There are now about 28,000 entries.

The show is not just about the Best Dog in the show but has competitions for dog agility and obedience. It also runs relay style races and judges young handlers. It also registers cross-breeds. There are a large number of trade stands selling everything to do with dogs from tinned food to dog beds.

The show was covered for years and televised by the BBC. Then in 2008 the BBC brought out a large criticism about the genetic diseases caused by years of in breeding. A controversy arose about the contorting and exaggerating of various characteristics such as short muzzles as in Pugs and Pekingese, and large forequarters as in Bulldogs. These characteristics pander to our love of neoteny or the retention of juvenile features in an adult animal. New health plans have now been formulated. Another recent story along the same lines is that of the Dalmatian which suffers from an underlying genetic problem with uric acid. This can lead to pain, gout, and early death. In the USA a healthy version has been bred by crossing Pointers with Dalmatians, which look the same, have the same white and black spots BUT neither the American nor the English Kennel Club will accept them. They seem to have lost sight of the fact that a Breed is a man-made concept moulded from the raw material of wolves.

But dogs have taken on many more roles in some parts of the world while being reviled in others. These plus the multitude of unexpected consequences will be seen in later chapters.

ETCETERA

1. Canadian Indians called the Milky Way "Wolf Trail".
2. The controversial philosopher Diogenes of Sinope (412-323BC) who famously lived in a barrel was called the Dog and when asked why, said "Because I fawn on those who give me anything and bark at those who give me nothing and bite the rogues." His followers were the Cynics, which word is derived from the Greek word for Dog. He maintained that man had become ridiculously artificial... and that was even back in ancient Greek times, so what would he have thought of the C21st? He strongly advocated a return to simplicity and cited dogs as role models as they slept anywhere, ate anything and performed their natural functions in public – all of which Diogenes also did. The word 'cynic' actually means dog-like.
3. Vasco de Balboa (1475-1519) a conquistador who crossed the Isthmus of Panama and the first to see the Pacific from the New World came across a tribe called the Quarequas and was so disgusted to find them performing homosexual acts that he set his dogs upon them and they were torn to bits.
4. The Beast of Gévaudan was a name applied to more than one animal in the mid 1700's in the Dordogne part of France. It was claimed to have killed over 200 people. King Louis XV sent professional wolf-hunters and 8 bloodhounds to hunt it. One wolf was eventually killed but more predations took place and people thought it a were-wolf or a cross breed with a dog or a hyena or even a mesonychid which is an extinct mammal with five hooves rather than claws.
5. North Korea's Kim Jong II – the 'Dear Leader' had 200,000 political prisoners and amongst other tortures and executions, dogs were trained to set on prisoners and maul them to death.
6. During WWII, Simon's Town on the Cape Peninsula was an essential port for the Royal Navy on their way east. The sailors would take the train into Cape Town, less than an hour away and there, deprived of liquor for so long, give way to drinking all they could, and then sometimes found it hard to find their way to the

station. Luckily for them help was at hand in the unlikely form of a large brindled Great Dane called Nuisance who had taken it upon himself to round up all lost and drunken sailors and escort them to Cape Town station, onto the train and back to Simon's Town. He never bothered with soldiers or airmen or even naval officers...only Jolly Jack Tars. Nuisance was made an Able Seaman and when he died was granted a naval burial.

7. The first living creature to orbit the earth was a dog named Laika (which means "barker" in Russian) a stray dog off the streets of Moscow; she was sent off aboard the Soviet Union's Sputnik 2 on November 3rd 1957. But as there was no way to bring her back, she was the first living creature to give her life to space travel.

8. Kipling in "The Just So Stories" told the tale of the woman throwing a roasted mutton bone to the Wild Dog who craved another one even though he considered man to be his enemy. The woman bargained "Wild Thing out of the Wild Woods, help my Man to hunt through the day and guard this cave at night and I will give you as many roast bones as you need."

9. Memorial to Lord Byron's gentle Newfoundland who died of rabies in 1808:-

> Near this spot
> Are deposited the Remains of one
> Who possessed Beauty without Vanity
> Strength without Insolence,
> Courage without Ferocity
> And all the Virtues of Man without his Vices.
> This Praise, which would be unmeaning
> Flattery
> If inscribed over human ashes,
> Is but a just tribute to the Memory of
> BOATSWAIN, a dog.

10. Argus was Odysseus' dog in Homer's Odyssey and when his master came home after years of adventure, dressed as a beggar, Argus recognised him.

11. King Charles II loved his small spaniels which became known as King Charles spaniels. By royal decree, which is still in the statutes books, they are allowed in any public place including the Houses of Parliament.

12. The Lascaux Caves with paintings from the Upper Palaeolithic (over 13,000 years ago) were discovered in 1940 by a dog called Robot owned by a young boy aged 17.

13. Smokey was a WWII mascot, a 4lb Yorkshire terrier owned by Corporal Bill Wynn. For two years he flew on combat missions in fighter planes and was taken on battlefields, where he comforted the wounded, making him probably the first therapy dog on record.

14. The Brown Dog Affair. A vivisection was being performed (under anaesthetic) by William Bayliss (1860 – 1924) the physiologist, who discovered hormones, on a brown dog. This triggered a political controversy and riots and marches in London by anti-vivisectionists at the same time as the suffragettes were out in force.

---oooO0Oooo---

CHAPTER 2

MEAT, MILK & MUCK

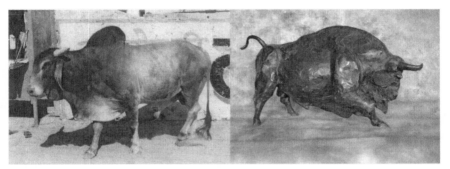

Zebu and Aurochs (sculpture by the author)

"Of the cattle are some for burden and some for meat:
eat what Allah hath provided for you, and follow not the
footsteps of Satan: for he is to you and avowed enemy."
Koran

The next animal we domesticated proved to be the first really useful addition to aid mobility worldwide. This was in about 6,500BC; but this creature was big – two metres at the shoulder, weighed a ton and was armed with formidable horns as can be seen in pictures from the Lascaux Caves. It was the Aurochs, the ancestor of all our domestic cattle. You only have to encounter a bison or better still an African buffalo to realise what Man was taking on and he was still barefoot and equipped with only puny weapons. Their horns, their size and their ferocity make you wonder at the audacity of Man converting them to domesticity and not just killing them.

Luckily, worship of the cattle family was an early response of Man to them and that gave the species a breathing space and saved it from the extinction we inflicted on so many animals. However once we had got together with them they turned out to be amazingly adapted for a range of conditions from yaks living in freezing conditions at 15,000 feet to the humped Indian zebu cattle and water buffalo in boiling hot swamps. Riding, ploughing and later pulling wagons, cattle took on many roles including providing milk and meat and manure.

The prehistory of cattle is confusing and continually being updated. How did cattle come to be not only in Europe but also in Asia and Africa? According to the Palaeontology Museum of Oslo the aurochs actually originally evolved in India probably some two million years ago. It then moved through the Middle East and Asia, reaching Europe roughly 250,000 years ago, by coincidence about the same time that *Homo sapiens* was developing. It may even be that two species of aurochs diverged hundreds of thousands years ago. Others think the gaur is the ancestor of Indian cattle, but the gaur is also related to the aurochs.

It seems it was only possible to domesticate the Eurasian/ North African versions of the cattle family. Domestication of the aurochs started in the southern Caucasus and northern Mesopotamia in about the 6th millennium BC and was probably domesticated independently in both N. Africa and India; the descendants have parted from each other in outstanding ways. India had its own cattle ancestor *Bos primigenius namadicus* which was possible the ancestor of the zebu; North Africa also had *Bos primigenius mauretanicus,* but again opinions differ on these ancestors. Linnaeus (1707-1778) originally classified cattle as three separate species the *Bos Taurus* or European cattle, the *Bos indicus* or zebu and the *Bos primigenius* but now they are grouped as one species, with sub-species. However, when dealing with events so long ago you can only approximate and must remain flexible as new data is turning up all the time. As I said, the history of cattle is very confusing.

Somehow, in spite of the kings and nobles hunting them almost to extinction, the ancestral aurochs together with the European bison continued to cling to existence in the forests of the Rhine right up until the C12th. The remnants struggled on and managed to survive until the early C15th in East Prussia and Lithuania and a few lingered on in Poland in the Jaktorowka Forest Park until 1627 when the last one, a female, unfortunately died. In Britain they had been wiped out even before the

Romans had arrived. (In 1998, the skeleton of an aurochs was discovered at Porlock Bay, on the edge of Exmoor.)

Caesar encountered them in the Gallic Wars and described them rather exaggeratedly "They are a little below the elephant in size and... their strength and speed are extraordinary. They spare neither man nor wild beast which they have espied." So when referring to the aurochs they were dubbed the 'Urus of Caesar'.

An interesting experiment was conducted between the two world wars by two brothers Professor Heinz Heck, Director of Zoo Park in Munich and Professor Lutz Heck of the Berlin Zoo who, encouraged by the Nazis, especially Goering, tried to breed back the extinct aurochs (*Bos primigenius*). They crossbred all sorts of cattle – Highland cattle, fighting bulls and others in a seemingly haphazard way and amazingly by 1932 had produced some good specimens, which even more surprisingly bred true. But modern geneticists do not believe you can resurrect an extinct breed and the experiment seems to have been conducted sloppily. The two brothers ended up with different specimens but neither resembled their ancient ancestor. They were not nearly as large, the bulls standing at 1.4m and weighing 600kg whereas the aurochs could stand at up to 2m and weighed about 1000kg. The aurochs had longer legs, longer neck and different horns. However it seems the Spanish Fighting Bull isolated from other cattle and bred selectively for aggression has an unusually ancient genetic pool and seems a very close match to the aurochs. Some of these pseudo aurochs were being kept in Devon in 2014 and showed one real aurochs trait of being most dangerous and difficult.

Of all the extant wild cattle, the gaur (*Bos gaurus*) is the largest and most distinguished with a high dorsal ridge and a huge dewlap of skin running from under the chin to between the forelegs. The males' coats are dark and they have white stockings; both sexes bear horns that are yellow with black tips curving upwards to 80 cm in length or more. Not unusually the males are much larger than the females and can weigh up to 1,500kilos and reach 2 metres at the shoulder and up to 4 metres long. When attacked they form a protective circle round the calves and the most vulnerable in the herd and they are quite capable of even taking on a tiger. The wild form is hugely threatened by the usual toxic combination of poaching and the endless illegal logging to turn the rainforest into pasture or for the growing of bio-fuels. However there are efforts being made to save it from being hunted into extinction. It is listed by the Convention on International Trade in Endangered Species (CITES) as an endangered species, and is protected in quite a few reserves such as the Royal Chitwan National Park in Nepal. There is also a new scheme by the Integrated Conservation Research to save both the gaur and the rainforest and elevate the local economy at the same time, as the gaur would produce a much better return than the poorly bred currently farmed cattle and also attract tourism as an alternative money-maker.

There are other wild cattle that have never been domesticated such as the African Buffalo (*Syncerus caffer*), the Musk Oxen (*Ovibus moschatus*), and the European Bison (*Bison bonasus*). Indonesia also has pygmy water buffalo called Anoa; there are two types, the lowland one (*Bubalus depressicornis*) and the mountain one (*Bubalus quarlesi*). Both have short straight horns and have quite a resemblance to deer. In Cambodia there is the Kouprey (*Bos sauveli)* which was only discovered in the thirties and may well now be extinct.

Another horror story of man versus wild animal concerns the American bison (*Bison bison*) which once numbered hundreds of thousands and was only saved on the brink of extinction. This animal is a semi exception of domestication as in 1965 after many attempts they managed to cross bison with domestic cattle and produce a creature, obviously for eating, which they called a Beefalo.

In England there are a few herds of white cattle of ancient lineage kept by various noblemen in their wild state and isolated from cross breeding, which are interesting to see. One such is the Chillingham herd kept at Alnwick in Northumberland, the seat of the Earl of Tankerville since the end of the C13th, which are not just white but have foxy red ears. They may possibly have been introduced by the Romans. The Chartley herd at Woburn Abbey in Bedfordshire was nearly wiped out by disease but being crossed with Longhorns led to their survival. There are also the Dynevor herd in Wales and the Cadzow herd in Scotland. These herds are also white but with black points. Seeing them gives you just a glimpse of the ancestral cattle of hundreds of years ago.

Cattle are ungulates, which mean they have hooves. They also have the special digestive system peculiar to 'ruminants', which means that they chew the cud. First they eat large quantities of rather indigestible grass and other raw matter that goes into the first of four stomachs – the rumen. The cud consists of regurgitated food that is then re-chewed and re-swallowed until it becomes digestible. This action is called ruminating, a term borrowed by Man to describe the mental action of pondering something over and over and over again. (Other ruminants are camels, giraffe, deer and antelope.)

Long before Man could start his interfering with cattle breeding, they had themselves turned into various very different animals – the zebu, the yak, the water buffalo and a selection of European cattle. Gradually cattle were being drawn into our lives but there were complications. There was settled life and there was nomadic life and the two did not always exist happily together. In about 10,000BC man had settled down in some places in Eurasia and started agriculture, and this led to the beginning of a form of village life. The very first draught use of cattle began in Mesopotamia in about 6,000 BC. This was before the wheel was invented so sledges were used. These early sledges were made from a "V" or "Y" shaped forked branch. The two trailing parts were joined with short pieces of timber to form a loading platform. The pointed end of the "Y" trunk was attached to the animals by a chain, making a type of travois. Traces of these travois drawn by oxen in a Neolithic village dating from 3,100BC have been discovered at Lake Chalain (Jura).

As perfect proof, a sledge-throne drawn by bovines was found at Ur, the pre-historic Mesopotamian port on the Persian Gulf, in the tomb of Queen Pu-Abi from the mid 3rd millennium BC. Also found were pottery vessels of Oxen with nose ropes as opposed to a proper bridle. The first real archaeological evidence of wheeled transport was found in Europe dated 2,500BC and with it, as absolute proof, were a pair of oxen skeletons.

Initially animals were yoked for traction in a most curious way – ropes were attached to their horns, for the simple reason that horns had always been considered sacred. And this inefficient harness was used for the first ox driven plough – a landmark of the utmost importance in Man's agricultural development.

However an early unexpected consequence of owning giant herds of cattle was forced mass migration in search of grazing. And this is where the settled villagers began to suffer. Apart from warfare, some of the early mass migrations were the result of ecological disasters such as drought, or denuding of the soil and certainly one of the main reasons for the invasion of the Indian sub-continent was the search for fresh grazing for cattle. And for these and other reasons sometimes settled communities returned to the nomadic way of life. This led to man propelled cattle invasions.

From 4,400BC nomadic herds and their herdsmen were moving across into Europe and Scandinavia with the result that they totally disrupted a peaceful and quite advanced farm culture. Cattle formed the capital of the Mongols, just as they came to do for many of the African tribes. Those early pastoral tribes such as the Cimerians (1,000BC) and the Iron Age Scythians (C8th BC) moved across the Asian steppes. The search for new pastures by about the C5th BC had evolved into the multi-use of animals for milk, breeding, draught, wool, and being able to put to use land that was too poor to grow crops but suitable for grazing. Then came the Sarmartians who were ascendant in the C2nd BC. In the C7th the Chinese Tang dynasty took control of Mongolia and encouraged the Mongols in internecine warfare. The next great event was the advent of the great Genghis Khan (1162 – 1227) under whose rule and under that of his grandson Kublai the hordes penetrated much of India and Persia and even China. The other most influential and famous were the Golden Hordes of Tamerlane (1336-1405). All these migrants carried with them not only their herds but their burgeoning cultures, arts, languages, religions, techniques as well as goods. The tribes on the move also spread their genes, their diseases and the diseases borne by their cattle. The Khans were also prolific breeders themselves and DNA attributes 16 million descendants to Genghiz, alone. Kublai Khan, also outstandingly virile, had 22 legitimate sons and is supposed to have added 30 virgins to his harem every year so he must have an army of descendants. But also these tribal wanderings could lead to clashes and raiding and then inevitably to warfare. The movements Eastwards took the Indo-European languages right across Asia to the East and the contrary movements brought the Turkic and Mongolian languages to the West.

Uses for cattle were becoming more and more varied. Eventually, apart from worship and myths and legends, man realised that cattle was becoming the most important animal we had domesticated up to then and had a multitude of practical uses – milk, meat, leather, horns and a still new concept – draught. The word 'cattle' is a collective noun. Herds are counted as "100 head of cattle", although some farmers in Australia, Scotland and Canada use the term 'cattlebeast' to denote a single animal.

Another attitude to cattle was that they were seen almost as a form of currency; certainly in Britain and Ireland they were so important to the economy that they were used in this way. The very word cattle derives from 'caput' = head and is related to 'chattel' = unit of property, also capital = property. For example, in Ireland in the Bronze Age, a slave woman was worth three cows! The word "pecuniary" is derived from the Latin 'pecus' meaning 'cow'. So presumably the expression "pecuniary advantage" must once have meant an extra-large herd of cattle. In rural Africa they were used in the place of money right into the middle of the C20th. And in many African tribes cows are still used in payment (*lobola*) from the parents of the groom to the parents of the bride. In the late C19th amongst the Masai you could get a wife for five beads, but a cow would cost you ten.

Milking was another matter. The earliest representation of milking in Sumerian times was found at Ur at the temple of Nin-Hursag and is dated at roughly 2,400BC. Ur had its own Bull God called Nanna. Milk was certainly consumed in Egypt and Mesopotamia but in certain parts of the world there was and still is no tradition of drinking milk. The ability in adults to absorb milk has had to evolve over a long period. Infants need milk, adults do not and the enzyme for lactase is switched off once infancy is past. 'Lactose mal absorption' is the name given to this inability for adults to digest milk and absorb lactose and is present in Central and West Africa, China, Malaya and the East Indies. However adults from early times could eat cheese and then we evolved the ability to digest milk which being an easily procured food for pastoralists, had the advantage of being germ free in places where the water is contaminated.

Cattle were being moved all over Eurasia by various tribes but breeding for purpose was not practiced until Roman times. It was they who introduced animal husbandry throughout their extensive empire and became the first people to really breed animals selectively, so it was under their programs that cattle began to improve in size and performance. When they invaded England in AD43, they took herds of cattle with them and gave the British that taste for beef which has remained with them ever since. They also discovered that castrating bulls produced oxen far more suitable in temperament for draught. Castration is a way of managing male animals by eradicating the animal's sex urges and turning aggression into submission. It can also have an effect on the physiognomy of the animal, for example with oxen the bones continue to grow and so this operation in cattle also leads to greater bulk and strength. They really valued their cattle more for draught purposes than for meat or milk and it was they who introduced a wooden yoke fastening across the shoulders of the ox so as to distribute the force and the drag weight. Oxen can pull harder and longer than horses and although slower they are also more sure-footed. Their advent transformed the pattern of farming and improved transport in an important leap forward. This means they really did contribute to our necessity for movement.

Sledges are still used in rural parts of East Africa such as Tanzania but when sledges are dragged across soft ground, they, of course, leave ruts, which when it rains can cause soil erosion. So in some countries such as Botswana they have been banned because of this erosion risk. I personally remember seeing a form of sledge drawn by cattle in Basutoland (Lesotho) as recently as the late 1950's.

The second step in the evolution of agricultural power was provided by oxen– first Man, then Ox, then Horse and finally the Tractor. For light work just two oxen were yoked together but for heavy work teams of fourteen or more were used. There are several advantages in using oxen as opposed to horses. For one thing their staying power, strength and temperament are superior for the job. And of course, not only are they considerably cheaper to buy and to feed but neither do they require as much rest or care. The main disadvantage is that they lack the horse's speed. Now oxen, as draught animals and beasts of burden have all but vanished from the roads of the western world except in Spain where they are still used to draw wagons for their numerous fiestas. However, in the developing world, Africa, India and Asia, they are still very much in evidence.

During the Middle Ages cattle again degenerated. The political conditions were chaotic and the art of breeding was temporarily lost under mismanagement by churches and monasteries and the rise of ignorance. The withers height, which under the Romans had reached 140cm, fell to under 100cm. But still they managed to have their orgies of beef eating which, as an extra excuse, were venerated for giving courage. Of course in those times, the guests at the table were arranged and fed according to rank and your portion of beef was matched to your status.

Beef eating in England became more and more popular until finally the greed for beef grew so bad that Edward II in 1283 passed a decree limiting the number of meat courses at a meal. Eating of beef became even more popular in England in the early C15th after plague had decimated the human population. The Beefeaters were formed in 1485 by Henry VII after the Battle of Bosworth, and the generally accepted reason for this nickname for the Yeoman Warders is that Beef was used as part of their pay, and this habit continued right up to the early C19th.

Once the human population was on the rise again, meatless days were introduced in the Roman Catholic calendar and not until the mid C19th was beef fully back on the rich man's menu.

Breeding better cattle up again was also on the menu. Man, of course had used selection to breed an amazing range of cattle, selected for milk, for beef, or for draught; for size and colour too. There are now over 400 species named. The Dutch aided by their rich polder land and good grass had bred some very good cattle and in the C17th and C18th they were imported into the British Isles and became the basis for several breeds. In 1799 the Smithfield Club was formed to show cattle and other farm animals and apparently the size of the farmers vied with the size of their cattle. By the tail end of the C18th cattle had doubled in weight. Smithfield Market, near the Barbican in Central London has traded meat and livestock for over a thousand years. The present Market Buildings were built in 1868.

The biggest cattle are the Holsteins-Friesians, well-known for their black and white patched colour, which can weigh 2,600lbs and even up to 3,000lbs and have been bred from 2,000years ago. The most famous bull, born in 1965, had the most amazing name of Round Oak Rag Apple Elevation. He sired 70,000 cattle and has over 5 million descendants and was named "Bull of the Century" in 1999 by the Holstein International Association.

Just to mention a few other renowned breeds – the Shorthorn breed originated in the North East of England in the late C18th mostly red of colour it was suitable for both beef and dairy until the late C20th when they were selectively bred as Beef Shorthorn and as Dairy Shorthorn. A very special example of this breed was the Durham Ox, born in 1796 and castrated. He grew into a magnificent animal and was sold by his breeder Charles Colling to a John Day for £250 who took him on tours round England to agricultural shows and even to London. The Ox, in a special cart, was drawn by four horses and earned his owner enormous amounts of money from the fascinated public. He was painted by several artists; when he died he weighed 1,200kg.

The English Longhorn was also an ancient breed with a distinctive white stripe down its back but not to be confused with the Texas Longhorn which was the first cattle to sweep across North America. The Texas version got there in several stages from Spain, first to the Canary Is in the 1400's, then to Hispaniola with Christopher Columbus on his second voyage in 1493 and finally to North America a few years later. Their horns can reach up to 7feet span. The early settlers interbred them with feral Mexican cattle and this provided a stalwart stock which ran wild in Mexico. Eventually they were replaced by more modern forms of cattle and remnants are now kept in a reserve.

The Aberdeen Angus was bred in Scotland 400 years ago. They are sometimes black and sometimes red and polled (without horns). They are essentially beef cattle and so good is their meat that McDonald's in Australia and USA have made a feature of special beef burgers – named Grand Angus and Mighty Angus. Other well-known and excellent beef cattle are Charolais which are one of the oldest French breeds, heavy, large and creamy-white.

The Italians have their own special Chianina cattle which is one of the oldest breeds, with a history going back over 2,000 years. Columella (AD4 – 70) who was one of the most influential writers on agriculture in the Roman Empire mentions them in Volume 6 of his famous farming books "De Rustica". The bulls could reach 2 metres high and weigh as much as 1,600kgs. They were used for meat and draught. At the other extreme is the Dexter which is half the size of a Charolais 90-120cm for cows and bulls. It is becoming popular for small acreage farms, as are also mini-Jerseys and mini-Herefords. There is even a Miniature Cattle Breed Society & Registry.

But the most grotesquely bred beef cattle must surely be the Belgian Blue in which a defective gene is used to promote the most enormous bulges of beef and muscle, passed on by artificial insemination technology.

In Europe Man has also bred special cows for their milk – Jersey c.1700, Guernsey, Brown-Swiss, and Ayrshire, to name a few and these dairy cattle have been selected down the years to grow enormous udders and produce great quantities of milk. Some, like the Jersey and Swiss, produce less but richer milk. From milk is made butter, cream, cheese, yogurt, ice cream and is used in cooking white sauces and puddings. Obviously to start milk production, a cow must first have a calf, which is removed within 24 hours. This sounds crueller than it is because it is essential and actually kinder to prevent mother and child bonding. The size of herd varies from country to country and farm to farm, with the USA having giant herds up to 800. The EU tried to have a set size and plan for cattle sheds, which when you compare wet windy, cold Shetland Islands with baking hot Spain seems another ridiculous example of one size simply not fitting all.

* * * * * * *

The East and Far East have three very special cattle – the WYZ of the cattle alphabet. The zebu (*Bos primigenius indicus*) is India's very own cow with its identifying fatty, muscular hump, droopy ears and large dewlap and is possibly the oldest form of domesticated cattle, from about 10,000 years ago. Cylinder seals from 2,500 – 1,500BC found in the Indus valley depict them and also in ancient Sumeria and Babylon. Remains of bones from 3,500BC were found in the Quetta valley in Baluchistan, in fact so prevalent were they in prehistoric times that their bones make up 80% of the collection from all archaeological sites. It forms a separate group from the European cattle.

As always, eating of beef was a matter of prestige and reserved for the upper class Brahmans. Almost like a rehearsal of the disaster that is developing now – by 300BC forests were being cut down, the land was becoming denuded. Meanwhile the human population was growing and spreading but the peasants *had* to have cattle to do the ploughing and provide milk and ghee. There was also the useful by-product of manure to fertilise the newly cultivated fields and to use as fuel. Worship was being replaced by work. Maybe this multitude of uses for the cow made the cow of much more importance alive and providing rather than dead and eaten. Then into this impasse arrived the new religion Buddhism in C5th BC that forbad the taking of any form of animal life and this led to a long struggle ending in the Hindus adopting the opposite precept of NOT eating beef themselves. From the Vedic period in India (2nd millennium – C7th BC) and in the Rigveda, one of the Hindu sacred books the killing of a cow became a capital offence. But although cow slaughter is still officially banned, there are rumoured to be at least 30,000 illegal slaughter houses in India. The meat is often served in kebabs and passed off as mutton.

Hindus venerate all cows and see them as members of the family, decorating them with garlands of flowers and tassels. Gandhi called the cow "a mother to millions of Indians". Today in India the ox provides traction for at least 60 million small farmers whose lands provide food for 80% of the population. "The cow is our mother, for she gives us milk'. But the Zebu's milk yield compared to European cattle is pathetic – less than a tenth at the best and quite often the Indian cows yield no milk at all. When a farmer delivers milk, the custom is to bring the actual cow to the door and milk them for the customer to prove that no adulteration has taken place. But their use as draught animals on small farms is far more essential and serves to avoid the peasants leaving the land to join the overcrowded city slums. Another very important by-product of cattle is manure and there is an incredible estimate of 700 million tons per annum. A good half goes on fertilizer and most of the rest provides cooking fuel. Some is used for flooring, as it is in Africa where it is mixed with blood to give a smooth hard shiny surface. The burning of cattle dung acts as an insecticide against mosquitoes. Another by-product is that the presence of manure on the land encourages dung beetles who take the dung below ground which improves the soil. They also make tunnels which encourage the earthworms and the ground is made more receptive to water – a marvellous example of a chain of links in nature. However on the minus side there are possibly 100 million useless cattle in India, animals that are too old to produce milk and of course they are not killed for meat. Many wander the streets like beggars looking for food, eating rubbish and plastic bags; they provide a terrible hazard for drivers as the punishment for killing a cow can be death. Although most Hindus are lacto-vegetarians quite a few of the lower caste including the Dalits (Untouchables) do eat meat – chicken, lamb and especially beef, which is half the price. Other beef eaters are the Muslims and the Christians. In fact only 40% of Indians are vegetarians out of a population of 1.2 billion. The cattle population is the world's biggest at over 300 million.

An unfortunate spin off of cattle being so valued in India for their milk and their sacred powers is that they are often given a drug for anti-inflammatory treatment that has turned out to have the most terrible side effect; it is deadly poisonous to vultures. These birds which can resist plague and botulism and rotting fly blown meat are totally unable to take 'diclofenac' even in the second hand form in the flesh of the consumer of the pills. As a result at least 97% of India's vultures have been wiped out with renal failure. This is bad enough generally but there is another side effect. The Parsee religion rules that their dead are placed on the top of tall towers called 'Towers of Silence'. The vultures then arrive to do their work and consume the corpses. It is called 'Burial in the Sky', but now there is a long queue of dead Parsees owing to the tragic vulture problem. Another side effect is that the populations of feral dogs have increased greatly increasing the risk of rabies. The drug was banned in 2005 but it will take some time for the reversal of the desperate effects.

There are 75 or more known breeds of zebu not only across south Asia but also in Africa. They were imported into Africa several thousand years ago, and having more sweat glands than other cattle they are well able to withstand heat and humidity. As they were found all along the East Coast and in Madagascar, they may have been introduced by sea when the settlers came from Indonesia in about AD500, or the Africans who came the relatively short distance from the African continent may have brought them back with them. As is well known they can multi-task and are used as draught oxen as well as for dairy and beef. They have been continuously imported into Africa and hybridized with the local taurine cattle, such as the Afrikaner. Another advantage is that they seem almost resistant to the rinderpest scourge, the cattle plague that kills so many animals and not only cattle but buffalo and even giraffe. It was pronounced eradicated in June 2011.

Zebus were also introduced into the United States about 1849 and by crossbreeding produced very successful hybrids. In Brazil they were crossbred with Charolais and this crossbreed is called Chanchim. They seem to be the only breed that can inhabit the tropical rain forest and resist all the parasites and diseases.

Yet another form of cattle, the Indian water buffalo (*Bubalus bubalis*) are also extremely widespread in Asia. They extended their range during the Pleistocene climate changes from Eastern Nepal and India, East to Vietnam and south to Malaysia moving out from India into NE & NW Manchuria. The wild version of the Water Buffalo is a massive animal and can weigh up to 2,600lb (1,200 kg) and has horns with the widest spread of any bovid – over six feet (2 metres) and can even take on a tiger by forming a line and performing a cavalry charge. They are now endangered and intermixed with other feral cattle, and the last estimate was that there are only 4,000 left, whereas there are 141 million domestic ones. It was probably originally domesticated in China and India in about 8,000BC and took about 3,000 years to reach the Near East. The curious thing is how long it took to reach Egypt; the Nile Valley was so obviously the perfect place for them.

But taming these huge creatures with their dangerous horns must have been a challenge of the highest order. Perhaps they were tempted with salt for licking and water to drink and bathe in. Their attraction to predators would have been another problem and their tendency to trample crops a further drawback but the prizes were worthwhile – milk, meat and pulling the plough. No wonder the Neolithic farmers tended to breed from the smaller more biddable specimens.

Their smaller domesticated descendants are ideal for working in swampy conditions and as more than half the world depends on rice, they are essential in the paddy fields and for transporting the rice to market. However although they are very strong, they cannot compete with horses or ordinary cattle as a draught animal. They can manage on very poor feeding but their main requirement is that they have to be able to wallow in water to overcome their reaction to hot weather. But on the other hand they cannot endure cold weather. There are two species, the Swamp one which is not fussy about where it wallows and the River one which prefers clean water. Strangely enough, Water Buffalo cannot interbreed with other domestic breeds of cattle, as neither can the African Buffalo.

In Italy the Water Buffalo's rich butter-high milk, which is made into ghee in the Far East, is used for making the delicious and famous Mozzarella cheese. There are, as always, different theories about how they arrived in Italy. The one most often put forward is that they were introduced into Italy and Bulgaria by returning Crusaders; Alexander von Humboldt (1769 – 1859) the geographer and explorer, supports this hypothesis. But still another claim suggests they were known in the time of Aristotle and even mentioned by him. Then there is the Lombard historian Paul the Deacon who wrote "Historia Langobardorum" c 790AD, who maintains that the Water Buffaloes were brought to Italy during the reign of King Aigulf in about 600AD, possibly a gift from the Khan of the Avars from the Caucasus. Anyway, however and whenever they arrived in Italy there are now an estimated 200,000 of them.

Another continent where cattle are invaluable is Africa. The tribes of East and Southern Africa have traditionally depended on their cattle which produce just about everything they need for survival, from milk and meat for food, dung for fertilizer and building material, including their skins being used for shields and clothes. As well the bones and horns provided tools and even the urine was used for disinfectant. And of course the usual extra – the possession of many good cattle gave the owner prestige. Cattle were traditionally used for 'lobola' or bride price. The Dinka people of the Sudan have cattle with huge horns and they refer to themselves as the 'Lords of Men' but the 'Slaves of Cattle'. Cattle, sheep and goats were present down to the north edge of the Serengeti in the 3rd millennium BC but it was another 2,000 years before cattle reached South Africa.

In South Africa, from $C16^{th}$ all European agricultural development and all mining was entirely dependant on oxen. There were three main strains by then as a result of crossbreeding between the imported cattle and the local breeds, which had the advantage of being mainly immune to disease. There was the large and bulky Afrikander the result of a cross between Dutch cattle and Khoikhoi (Hottentot). It was rather subject to congenital goitre, but it had great stamina as well as being disease resistant. Then there was the Bechuane, a deep red colour, which was resistant to thirst. And finally the indigo Zulu or Nguni ox, which was speckled, hardy, relatively speedier but took lighter loads. The dazzling multi-coloured patterns on its skin give rise to fanciful names. The Xhosas, the Zulus and the Swazis brought them down into South Africa when they migrated south between 600 and 700AD. The importance of a village was measured by how many cattle they owned. King Shaka of the Zulus realising this, seized control of all the Nguni herds. Then he bred them according to skin colours, such as brown speckled, black flecked – a sort of designer skin grown for the uniforms. Each pattern was assigned to a specific regiment (or impi) and used for making loincloths and cloaks to match. Shaka's elite personal guard were clad in pure white.

Cattle were always venerated by the Africans but their herds were small and often nomadic. When colonial times came the nomadic life was discouraged by arbitrary fencing and colonial disapproval through the administrators of government services. Wells were drilled to encourage settlement, the wear and tear of cattle going to the water holes and overgrazing on a grand scale has resulted in desertification, yet another example of trouble in nature resulting from interference by Man.

Another important historic use of oxen in South Africa was the Great Trek in the 1830s and 1840s when about 12,000 Voortrekkers consisting of the original Dutch and Huguenot settlers moved away from the Cape when the English took it over. This great migration was contemporaneous with the opening up of America. The Afrikaaners had been all but isolated for a hundred years and had developed their own variations of religion, culture and language. They had mountains to cross but no roads. So when descending steep inclines they invented the cunning device of removing the back wheels of the wagon and substituting thick bushes to act as a kind of brake. When camping, for security, the ox wagons were drawn up in a circle.

A country that has many types of cattle is Nepal; they are nearly all humped, and are used for all the usual productions of milk, draught and ploughing. Rather unusually the oxen are shod for use on hard roads. A pair can draw a load weighing 1 ton in a two wheeler cart with iron tyres for the distance of 32kms in a day. In west Nepal they raise one of the smallest breeds in the world – the Achhami, named after the district where it lives. The bulls are less than 1 metre at the withers, have a great big fatty hump and only weigh 110 – 150kg, but their dwarf size is an asset in this hilly country. The cows give very good and amazingly plentiful milk.

Nepal is also the home of a very unusual and very different branch of the cattle family – the Yak. The wild form is *Bos mutus*, which is native to the high mountains of Tibet, Ladakh and Himachal Pradah or the Qinghai-Tibetan plateau, the highest continuous ecosystem in the world. These wild ones, of which there are but a few hundred left, are of course much bigger than the domesticated ones, in fact nearly double the size and the bulls can reach to as much as 230cm at the withers and weigh 1,000Kg. They were hunted almost to extinction but thanks to recent and more stringent conservation practice they are now staging a come back.

The domestic form of yak (*Bos grunniens*) or (*Poephagus grunniens*) is much smaller, just over half the size, but is of paramount importance to the people of the high Himalayas as they are adapted to the most extreme conditions of the cold, high terrain and the rough steep mountain paths. Yaks were and still are the main means of transport in the Himalayas. Without their resistance to high altitudes and desperate cold temperatures dropping as low as −40°C (the temperature that killed off so many of Napoleon's army on the retreat from Moscow), the Tibetans and the Nepalese could not have conducted the Trans Himalayan trade. The reason they can stand such extreme cold is thanks to being insulated by the combination of their long shaggy outer coats and thick matted undercoat. They make the most amazing beasts of burden as they can travel 20-30 km a day carrying a load of 330lbs or 150 kilos, thereby earning their nickname of "Ships of the Plateau". They are one of the few animals able to cope with altitudes above 14,000 ft and even up to 20,000 ft. thanks to their great lung capacity. Even their blood cells, being three times more numerous are specially adapted to carry oxygen. There are many different theories as to how many years ago they were domesticated – claims vary from 8,000BC, almost as long ago as pigs, down to 3,000BC – quite a time difference. The trouble is that it has been incredibly difficult to perform any archaeological research into their ancestry. There are several reasons for this. It has always been extremely difficult to gain access to Tibet quite apart from the actual logistics of travel to what was an extremely remote region. Many of the former Dalai Lamas actively discouraged foreigners visiting and poking about in the countryside as Buddhism, for religious reasons, is dead against excavations and even when the remains of dead yaks have been found it is almost impossible to distinguish the domesticated ones from wild ones. Also saddles and bridles which would have helped with dating were made of wood and leather and would have rotted away. And now foreigners have been excluded from Tibet since the annexation of the country by the Chinese took place (1949-1951). However what little research has been managed shows that although yaks are *bovids*, they are neither closely related to the aurochs or the zebu. This is according to the latest on interspecies relations, an imperfect and forever changing science.

The domesticating of them was quite a challenge as the wild ones are very fierce and fleet of foot. Life used to revolve round yak herding in Afghanistan and Tibet; also with the Hunza in Pakistan where they herd the yaks up to the summer pastures of the Karakoram. Yaks are amazingly efficient food-converters and like reindeer can subsist on coarse grass, lichen, and thorn-bushes and crunch snow and ice for liquid. Yak breeding has a great cachet, it is considered a noble occupation and so men will endure great hardship tending their herds. In fact the majority of yaks are owned by a relatively few wealthy families.

They are so incredibly sure footed they are able to detect crevasses under the snow. Apparently if the snow is too deep for the yaks to reach the grass they roll down the slope, dislodging the snow and then slowly eat their way up again. If a yak falls into a swamp, unlike horses or mules, they do not panic but use their legs like oars dragging their bodies rather like rafts across the mud. A cord is attached to the nose supposedly to guide it but it seems to go its own way, with its muzzle close to the ground and breathing hard and noisily. It has the most amazing balance and will jump chasms, glissade down gravelly slopes and show no fear of heights. But it can be very stubborn and can suddenly stop for no seeming reason and then refuse to budge.

The first clue to their difference from other cattle is in their very name – '*Bos grunniens*' means grunting like a pig, because yaks are unable to make the usual cow noises such as lowing and mooing. Even Linnaeus (1776) used this Latin word for grunting in his "Systema Naturae" and one of their nicknames is 'pig-like cattle'. Their other nickname is 'horse-tailed' cattle as they differ markedly at their rear end growing the most amazing long bushy tails, horse like, only three times more luxuriant. These amazing tails first reached the European markets during the reign of the Emperor Domitian, in the C1st; Lamaseries covet them and use them as flywhisks; in India the tails are called chowris; they are sometimes used as great flowing and very ornamental earrings for elephants; in China they dye them red. Unfortunately so much monastery art has been destroyed since the Cultural Revolution but luckily there is a magnificent fresco in the East Hall of Sakya Monastery (built 1,073AD) depicting several thousand labourers and a train of yaks bearing building materials.

Yaks provide just about everything to sustain life for Man. Yak hooves are extremely durable and much tougher than horses so their primary function is to provide transport and to thresh grain. But they also produce meat, milk and dung for fuel. The meat is not of very high quality as it lacks fat, but it is still the main source of protein for the Tibetans. But as the Buddhist religion prevents them from killing animals themselves they employ Muslims as butchers. The yaks' marvellous outer hair is woven into tents or yurts and also made into huge screens and drapes for the monasteries and the Potala Palace. Women even braid it into their hair to lengthen it. The wool is made into every kind of clothing, including knee high woollen boots. Their butter, called ghee, is not only eaten but also used as light and fuel for butter lamps, and even for paying wages. The leather is invaluable for clothes, boots and is also made into coracles for riding the rivers. Even the blood of yaks is valued for its supposed medicinal value, and sometimes fried or boiled as a food. The bones are used for tools and artifacts and their marvelous horns for decoration. As though that isn't enough, they can find their way and are so sure-footed that to open a mountain pass, choked with snow, you only have to drive a herd of yaks to and fro to trample a pathway. And finally they even qualify in the games arena as being used for a curious form of polo in Nepal, where it also appears on the five rupee note.

The Greeks had heard of yaks and named them "poiphagos", which means eater of grass. However it was the rather unreliable Marco Polo who was the first European to see yaks in the second half of C13th in a city he called Singui in Tangut. It was more than likely Si-ning, a town on the route between Lhasa and Peking and Shanghai. Marco Polo with his propensity for dramatic exaggeration described them as 'wild cattle like elephants'. They obviously were very numerous as in 1898 an explorer called Welby visiting Tibet made the comment that on "one green hill more yak visible than hill."

The Sherpas call the male yak "Yak" and the females "Nak" or "Dri" and there are more than ten breeds of domesticated yak and an estimated 12 million in the highlands of central Asia. In Nepal yaks are called Bhotey or Bhotia, after the Nepalese name for Tibet. The Sherpas of Nepal came there from Tibet about 3-400 years ago and live well above 3,000m and even up to 5,000m in the summer. At these altitudes the ground is frozen for half the year, which means the herds have to be constantly on the move seeking the right pasture at the right time of year. At night they are tied to pegs and guarded by Tibetan mastiffs.

Originally it was thought that yaks were really not able to endure being at altitudes lower than 10,000 feet as they were thought to be prone to disease at the lower altitude. There is a Tibetan proverb "yak will not live where maize can be grown" – meaning where it is too low. In spite of this, Warren Hastings 1st Governor General of India (1773 – 1785) took a bull back to England with him with certain success, and there are about 12-14 million domesticated yaks in China alone. Yaks are now being bred in the USA in Yampa Valley in Colorado in the Rockies. This valley is considerably lower than anything in the Himalayas being only about 6,400 ft. Even the highest peak Mt Elbert is only 14,433ft, but the yaks seem to be thriving. They are bred for their hair, particularly the soft undercoat, and the softest down, which can be compared favourably to cashmere.

Hybrids are bred by crossing male Yaks with domestic Zebu cows and are called Dzos. They show typical hybrid vigour, being nearly one fifth taller and 50% heavier and showing a more amenable temperament so they are very much valued as beasts of burden. The call they make is also a hybrid, a sort of cross between grunting and lowing. The cows yield more milk than the yaks. They are popular in Ladakh and Kashmir as Dzos can with stand warmer temperatures and the lower altitudes.

Not only are the horns quite common as decorations on house roofs but they have another unique but rather grisly use. There is a caste of Tibetans called Ragyabas, the equivalent of the Untouchables of India. Some of them are employed to break up the bodies of the dead and pulverize the bones and flesh together and then feed the vultures at a funeral site. For this they use yak horns as tools. So yaks provide the inhabitants of the

Himalayas with almost everything they need – food, transport, hair and skins, horns, highly prized tails and even food for funereal vultures.

* * * * * * * *

Meanwhile one aspect of cattle began to outstrip all others – beef. The hunger for beef in the western world has grown and grown. Man propelled cattle invasions began taking place. When the English began running out of pastureland they looked to Ireland and forced the disenfranchised Irish to farm on smaller and smaller plots, with their bigger and bigger families. In turn the Irish fatally turned to a monoculture – potatoes, and the Potato Blight in the mid C19th caused a terrible but well known famine and exodus. This poisoned relations between England and Ireland and we are still feeling the results.

The next candidate for providing beef and pasture was North America. After the Reconquista in Spain the nobles were granted large tracts of the land recovered from the Moors and employed this land for an early form of ranching sheep and cattle. These techniques travelled to the New World when Christopher Columbus took the first cattle (and horse) on his epic voyage. As a matter of fact one of the main if rather unglamorous reasons for this voyage to find a new route to the Far East was to bring back the spices that are used to preserve beef. The breed he took with him was the Spanish Longhorn, a very tough animal which was highly adaptive and became browsers as well as grazers. The Vaquero tradition of North Mexico adapted to the larger regions. Animals were branded for identifications purposes but then turned loose to forage for themselves.

Another unusual place for cattle ranching became Hawaii. They were introduced there by Captain George Vancouver (1757-1798) who was a great explorer and had previously sailed with Captain Cook. He presented several head of cattle to the Hawaiian King Kamehameha in 1793 and again in 1794. He instructed the king to turn them free to breed and place a ten year 'kapu' on them for protection. But this protection order was not rescinded until 1830 and bred they had. In spite of hunting now being allowed, by 1846 there were over 25,000 wild cattle and 10,000 domestic. The wild cattle trampled crops, ate the thatch off houses and wrought every sort of havoc. The Hawaiians couldn't contend with this plague on their own and ex-convicts from Botany Bay were hired to help with the hunting and culling.

Meanwhile the west was being virtually colonized by cattle barons and the native people were subjugated. The herds grew and grew into the millions and the new owners, the Beef Barons were the wealthy elite. There were just two awkward blocks – the Bison and the Amerindians who depended on them. But that was solved by the most horrific wholesale slaughter of a single species by man the world has ever seen. The buffalo which had been there for at least fifteen thousand years vanished almost over night – over 4 million of them were slaughtered. And the Indians couldn't believe the herds were gone: they never really recovered from the shock: their world was shattered. Most of the dead buffaloes were left to rot, but sometimes the bones were collected for fertilizer and the horns for buttons, glue and combs. By the end of the C19[th] an area the size of the whole of Western Europe had been turned into a giant pastureland.

The men in charge of these vast herds in the USA were called cowboys and they have achieved almost mythical status in American culture. A mass of fairy stories or folk legends have accrued about the frontier being pushed back with an almost religious drive and of man's dominion over beast and heroic battles with the 'Red Indians'. Films galore have been made in which cowboys are invariably portrayed as dashing heroic and very macho men. The truth is that they were usually poor landless men, certainly in Chile where they were known as 'hauso' or in the Argentine 'gaucho', 'llanero' in Venezuela, 'vaquero' in Mexico. In Hawaii cowboys were called 'paniola' which was the Hawaiian way of saying 'español'. Even the battles with the Amerindians rarely took place. And with the shortage of women, homosexual relationships were prevalent.

Reasons for eating Beef on this mammoth scale? Well apart from being energizing delicious food, it continued to signal wealth, status, and machismo. It indicated you had arrived in the top billion of those people who can afford to overeat, if they have a mind to. After the end of the American Civil War this beef eating addiction grew and grew. The growing herds of cattle were often owned by absentees – British financiers and aristocrats who were greedy for beef, not only the ultimate luxury of the upper class diet but a great investment.

Then Philip Armour opened a meat packing plant in Chicago. Texas was filled with cattle, but the next problem was to get them to market or rather the Chicago abattoirs. Jesse Chisholm opened a wagon trail in 1864 which became known as the Chisholm Trail nearly 1,000 miles in length, along which passed millions of cattle from Texas to Abilene to catch the trains onwards to Chicago. A balance had to be struck between driving them too fast for the cattle to maintain weight and health. The best average speed seemed to be about 15 miles a day, which could mean up to 2 months on the trail.

The next innovations in the late C19th that took over from the west were the railways and then came refrigerated cars. With the new refrigeration, driving of cattle became a thing of the past and from 1875 onwards carcasses were shipped to England. To fatten them up cattle were increasingly fed on grain which is not their natural food and could upset their digestive systems. To enclose the vast land seizures a new simple but effective invention by Joseph Glidden (1874) was employed – barbed wire – which was nicknamed 'the devil's hatband' by the cowboys. The days of free roaming were over.

The next continent ripe for plunder for 'free grass' was South America. In Brazil, Mexico and the Argentine, cattle were at first kept by the Jesuits at their missions. Here the barriers were the Rain Forests and the rural indigenous population. Money seemed to be no object so they began to cut down the forests and ignored the plight of the peasants.

Then it was Australia's turn. Australia's farms were mainly for sheep and wheat well into the late C19th. The cliché runs "Australia rode to prosperity on the sheep's back". The introduction of railways meant more opening up and accessibility for the remote farms in the outback. Then cattle began to be introduced in the early C19th. Water buffaloes were introduced to the Northern Territory of Australia in the 1820's and 30's as draught animals. Like so many introduced animals they became feral when the outposts were abandoned or modernised. They are still flourishing in spite of attempts to limit their numbers. Farming cattle is now big industry in Australia and they are kept on vast farms called cattle stations found in the 'outback'. The old-fashioned way of rounding up the vast herds of cattle by horse took weeks and is now supplanted by the use of special low-flying helicopters. The helicopters are flown dangerously low and slow in what is known as the 'dead man's zone'. Much of this beef is destined for the Far East.

These ranches or stations as they are called in Australia were mostly family businesses and some are the largest in the world. The farmers are called graziers and their cowboys are known as jackaroos. About 90% of the cattle are beef cattle, raised mainly in Queensland and NSW. Most of the dairy farms are in Victoria. The Cattle Council of Australia is twenty years old. Some of the early cattle were set free and became very successful feral animals adding to the long list of animals that Australia wishes had not been launched into her countryside. However the 'scrub bulls' provide some exciting hunting for the daring.

Movement in cattle started with pastoralists moving across Asia seeking grass and carrying their culture with them and has ended up with cattle being shipped all over the world seeking new pastures and being fed with grain to provide beef for the aspiring. It has been a very long haul from Astarte and Hathor in the heavens to hamburgers in MacDonald's. Most changes in our worship of cattle have definitely not been for the better.

Cattle drives were an important part of the economy of the American West; from 1866-1886 over 20 million were moved from Texas across Oklahoma to the railheads in Kansas for shipments to the stockyards in Chicago. Probably the most famous was the Chisholm Trail, about 520 miles long and as the herds covered only about 10-12 miles a day it took a long time.

ETCETERA

1. Best cuts of beef or venison were served at the high table at mediaeval feasts and the less desirable were served to those further down the table until you came to the serving of venison entrails which were called 'umbles'...hence the expression 'to eat humble pie'.

2. An interesting legacy of the Norman invasion of England in 1066 was that they obviously brought their own language and the fall-out from that is that the English words cow, calf and sheep are

used for the animals in the field and beef (boeuf) and veal (veau) and mutton (mouton) are used for the table.

3.   Hindu mothers bathe their sick children in cow's urine as a cure for certain diseases.

4.   Captain Alonso de Léon who led an expedition across northern Mexico in the late C17[th], is credited to having left a bull, a cow, a stallion and a mare at each river crossing. Some of these animals became feral.

5.   In 1992 an exciting discovery of an unknown member of the cattle family was made on the border between Vietnam and Cambodia. Unknown tiny insects are being discovered all the time but to find a new sizeable mammal was really amazing. This creature is the Saola or Vu Quang Ox (*Pseudoryx nghetinhensis*) and was discovered in the mountains on the borders of Laos and Vietnam. It has long straight horns like an oryx and the most unusual white face markings in spots and stripes and a deep chestnut body. It is quite small (100kg, 90cm at shoulder) and has a hunched back and compact neck and shoulders, which help it push its way through thick forest.

6.   One great example of the importance of beef cattle was a shorthorn castrated bull called "The Durham Ox". This was an early example of selective breeding by a Charles Colling at Brafferton in NE England in 1796. Weighing 3,000lbs, it was paraded round England in a special carriage drawn by four horses to agricultural fairs for over five years. It drew the crowds and the its fame was such that it was painted by several artists and china and porcelain figures made, including the famous Staffordshire blue and white china. And as a true sign of popularity several pubs were named after it.

7.   Beautiful Pictish carvings in stones have been found at an impressive Iron Age fortress on the Moray Firth. The best one of six stones out of an original thirty found in the C19[th] is called the Burghead Bull carved into the stone like a drawing with stylised scrolling to indicate the muscles.

8.   The constellation Taurus resembles a bull, which is one of the signs of the Zodiac.

9. Certain ants keep aphids as the equivalent of cows and 'milk' them for their honey dew excretions. To keep aphids handy, ants have a chemical on their feet that tranquillises them, sometimes they bit off the aphids' wings, and some keep them in pens underground.

10. Taboos over food especially beef have a long and bloody history. Rumours of use of cow and pork fat used to grease cartridges was one of the main causes of the Indian Mutiny.

11. In Delhi and Mumbai gangs of thieves steal cows at night and sell them on the Black Market for £250

---oooO0Oooo---

CHAPTER 3

EQUINES

Przwalski Horse and Heavy Horse

*"There is no secret so close as that between a rider and
his horse"*
Robert Surtees 1803-1864

*"The talk slid and the talk slid south
With the sliding puffs from the hookah mouth
Four things greater than all things are –
Women & Horses & Power & War."*
'Ballad of the King's Jester' by Kipling 1865-1936

*"When I bestride him, I soar, I am a hawk; he trots the
air; the earth sings when he touches it; the basest horn of
his hoof is more musical than the pipe of Hercules"*
William Shakespeare Henry V.

The next great love affair was between Man and Equines, in particular the Horse – the Prince of animals, the 'drinkers of the wind' – the creature to whom we owe the most for dragging us out of pre-history and into modern times. His story is full of remarkable incidents, near misses and amazing triumphs and it very, very nearly didn't take place. Horses were saved in Asia from extinction by a hair's breadth, only 6,000 years ago. The reason for their near extermination? We were eating them. Luckily better uses, as will be demonstrated, were found for them, just in time. This new animal that had galloped into our lives transformed them. In no time we were moving at what must have seemed astounding speed.

The fossil record of the horse is amazingly full and provides most compelling evidence in support of Evolution. Palaeontologists have been able to piece together a more complete picture of the development of the modern horse than of any other mammal species. These records are to be found in the continent where over a third of the population are Creationists and believe, like Bishop Ussher in the C17[th], that the world is only 6,000 years old.

Horses originated in North America approximately 52 million years ago. They co-existed with woolly mammoths and woolly rhinos as did reindeer and bison. Horses are *perissodactyls,* which is the term for an ungulate with an odd number of toes. The others in this order are tapirs and rhinos. Cattle, sheep, goats and pigs are ungulates with an even number of toes. The evolution of the fast-moving, single-hoofed beautiful creature we know as a horse from the early 'Dawn Horse' (*Eohippus)* is so fascinating that it is worth introducing some of the intermediates along the way. These early specimens whose remains have been found in North America weighed about 20kgs and stood about 40cm tall. They were rather like dogs but with long faces, which could accommodate their 44 teeth; another name for them was *Hyrocotherium* which gives a clue to their appearance meaning 'like a hyrax'. Another move in the right direction was that the original five digits were down to four on the front feet and three on the back. But still at this stage it was not remotely like the horse we know.

Moving on quite a few generations from Dawn Horse to *Mesohippus* in the early Oligocene era (34 million years ago to 23 million years ago) we now have an animal 24ins (61cm) at the shoulder, back less arched and legs getting longer. The toes were beginning to disappear and now consisted of a large middle hoof and two smaller. The next horse ancestor was Miohippus which existed for a considerable period in the later Oligocene era. Then on down to *Merychippus* which 18 million years ago was getting bigger still at 10 hands (40ins) and had shed the extra hooves and now had one per leg. Evolution was at work as always, and he now had a bristly mane and special teeth, which grew as they wore down, providing self-renewing grinders. This was essential, as they were no longer folivores (leaf eaters) of the forest but were grazing on the newly arrived grass of the savannah, or prairies. Part of the reason for the change in toes/hooves was that that the climate had changed and swamps had dried into prairies. These early horses had learned to survive on a low-quality fibrous diet, so the search for pasture caused them to roam over a vast territory. Grass has far fewer nutrients than leaves, so has to be consumed in vast quantities and that wears down the teeth. The rapid modification of horses' teeth is an example of evolution working at an accelerated rate. The branching of these horse ancestors into at least 19 new species was termed the "merychippine radiation" and this was the time of the greatest diversity and numbers of *equines* ever. Herds roamed all over the Americas. These migrations were reminiscent of the more recent herds of Bison in North America and as large as the wildebeest migration in Africa. It was the biggest success story – for a time. Just to link us up with our future workmates, our ancestors the great apes were emerging only 15 million years ago, and we shared a common ancestor with the chimpanzee only 6 million years ago – so in terms of evolution we are very junior to the horse.

After more changes down the millennia, about 4-5 million years ago, the real *equus Cabullus* evolved, still represented by at least three species. They had zebra-like bodies, striped legs and bristly manes. Then about 136,000 BC came one of the most extraordinary and inexplicable phenomena of all times – these modern forms of horses started moving out of America and over to Asia, not all at once of course, but gradually. This migration was enabled by the Beringia Land Bridge joining Asia and Canada, which was 2,000 miles long, longer even than Alaska today. At that time there was one fifth more land in the world than there is now. The great question is just why did the ancestor of the horse move from Colorado to Siberia? No one knows. But it was fortunate it did as it was doomed on the American continent, where they were wiped out approximately 11,000 years ago.

Homo sapiens had evolved about 500,000 years ago and they had moved the opposite way, also across the Beringia Land Bridge but from Asia into North America, in about 48,000 to 45,000 BC. (Also moving across the land bridge from Asia were herds of *Bison priscus*.) The men were of a Mongolian type and they travelled the whole length of America from North to South. Climate change in the late Pleistocene had seen the extinctions of large mammals in the Americas such as mammoths and sabre tooth tigers. So with the combination of this and overhunting by man, not surprisingly by 8,000BC the remaining horses were now totally extinct in the Western hemisphere. This means that horses are only known by the immense range of fossil remains in North America and they do not feature in any cave art. Recently a skeleton of an extinct horse with butcher marks and spearheads from the Clovis period (9,000 – 10,000 BC) was unearthed near Calgary, showing that the connection between man and horse, as suspected, was on the hunting and eating level. Men only reached Tierra de Fuego, about 6,000 BC, and a few remnants of wild horses might possibly have remained in this remote area despite the general migration north and west.

More importantly horses moved gradually across Eurasia and into Africa, but in nothing like the quantity and diversity of the merychippine radiation. It is generally thought that the ancestors of the horse had split from those of the zebra and the donkey between 4 million and two million years ago. So four related forms of *Equus* had by then developed:

1. Zebras in Sub Saharan Africa
2. Onagers or Hemiones in the Middle East
3. Asses or Donkeys in North Africa.
4. Horses in Europe and Western Asia.

From the point of view of domestication becoming the final judgement of man, we start with the most useless of the equines to that end – the zebras and quaggas. Like the examples we had in the cattle world – buffalo and bison, these were the equines that defied domestication. The most primitive living *equine* is the zebra, particularly the Grévy zebra which is related directly to the ancient equines whereas the other zebras are closer to modern asses and horses. The Grévy was named after the French president who was given one in 1882 by the Governor of Abyssinia, which he kept in the Ménagerie du Jardin des Plantes. The quagga (a sub-specie of the plains zebra), whose name came from the Hottentot dialect, has been extinct since 1883 when it was wiped out by the Dutch and English settlers, which was a pity in more than ways than one, as from all accounts, it was more tractable than other zebras and therefore might have been tameable and that would have made them very useful, as of course, it would have been resistant to the African diseases that decimated the imported European horses.

Of the other zebras, the Mountain Zebra was totally beyond discipline but the early Dutch did manage to train Burchell's Zebras to guard their horses from hyenas, and in the mid-C19th commercial firms in the Orange Free State such as the Zeederberg Transport Co. trained mixed teams of quagga and zebra and horses to pull their wagons, but the zebra never really became domesticated like the horse. There was no doubt they had their good points, the main one being their immunity to horse sickness, carried by the tsetse fly and fatal to so many horses in the early days of colonizing South Africa. But they can be very vicious and have the nasty habit of biting their handlers and even, on occasion of killing them. They also are very tricky to lasso as they have far better peripheral vision than horses. In the C18th according to R.J. Gordon (descended from Gordon of the Cape Garrison) zebras as draught animals were 'tough, good pullers but kicked and bit too much.'

However there were several amusing exceptions, typical of the mad early colonial times. Lionel Walter, the 2nd Baron Rothschild, the celebrated naturalist, actually managed to drive zebras in a four-in-hand carriage in London in the early 1900's. Another 'celebrity' was Josephine Dale Lace a Johannesburg socialite and one time mistress of Edward VII who, when she wasn't lying in a marble bath of milk, was driving a four-in-hand zebra cart. Then at the beginning of the twentieth century a Dr Rosendo Ribeiro, another eccentric, (the man who diagnosed bubonic plague), rode a zebra into Nairobi. And Elspeth Huxley, growing up in Africa relates how a friend of hers, Jock Cameron, after many failures, also managed to train four zebras, which he had raised from foals, to pull a cart. He was the only one who had any control over them and then not much! They fought each other, rushed about at full gallop or refused to move at all. Another couple of stories illustrate the unsuitability of taming zebras. One incident involved a dragoon guardsman who had planned to ride a zebra with the panache of a cowboy on a bucking bronco. But he was soon thrown off, dragged along the ground and had one ear bitten off by his maddened steed. So, not surprisingly, after a couple of centuries the domestication of the zebra was abandoned. All the zebra family except the Plains zebra are now endangered. By 1910 the trigger-happy early colonists had decimated their numbers.

One of the first draught equine animals to be really domesticated was the onager, or half-ass, about 5,000 years ago. Onager (*Equus hermionus*) is the Latin name for a wild ass or *Ane sauvage* or Asiatic Wild Ass. It is larger than a donkey but with very short legs. According to the Mediaeval Bestiary it was supposed to be able to predict the coming of the equinox. It also had the reputation of trying to castrate male foals, out of jealousy, according to legend, (Pliny the Elder C1[st]AD and Isodore of Seville C 7[th]AD). More likely, like lions, they were trying to kill competing breeding males. Their astonishing speed is quoted in the Koran and compared to the retreat of unbelievers *"as if they were asses fleeing before a lion"*, (Sura 74, Verse 50). Sumerian art of the 3rd millennium shows two and four wheeled vehicles being pulled by four onagers abreast – an early form of quadriga, so they were not totally untameable as was reputed. They had reached Mesopotamia by the 2nd millennium BC and were used by the Hittites. Herds of them were last recorded between the Tigris and the Euphrates sometime after 1845. They were re-introduced into Israel in the 1980's

Another wild half-ass, the one most like a horse, the Kiang (Tibetan name Djang) is endemic to the Tibetan plateau. It lives at altitudes above 13,000 feet and owing to its double length winter coat can endure temperatures as low as – 40° (the temperature that froze to death so many of Napoleon's army, including the horses on the retreat from Moscow). They are probably the half-ass half-horse mentioned in Pliny and Herodotus. Some of these, especially in Sikkim, are extremely endangered. More exist in Ladakh, Tibet and in China several thousand roam the Arjin Mountains Nature Reserve. These are very attractive creatures with rich reddish-brown on the upper parts of their bodies and pure white underneath with a dark chocolate coloured dorsal stripe down their backs, from the stiff upright mane to the end of the tail, which ends in a tuft of blackish brown hairs. They are the largest of the asses, some growing as much as 41 inches (10.1hands) at the shoulder. They have a matriarchal society, not unlike elephants, so it is the old females who lead herds of up to 400, while the males remain solitary, often running in bachelor groups until the mating season brings them together. They have never been domesticated.

However the ancestor of the donkey was neither of these Asian varieties – it was the African wild ass (*Equus africanus.)* The ass that least resembles a horse, this *equid* was indigenous to the African continent and had various subspecies scattered from the Red Sea in the East and the Atlas Mountains in the West, Ethiopia, and Somalia and possibly as far south as the Northern borders of Kenya. A few still live in the Danakil depression in Ethiopia. In the C16th – C17th asses were protected by the Sultan of Aussa or Afar (part of Ethiopia) with the strong deterrent that any man who killed one would have his hand cut off. The Nubian wild ass is now probably extinct – a great pity as it was described as a creature of splendour and beauty – terms you would hardly apply to the domestic donkey of today. The Somali Ass is also extremely attractive and luckily has survived and is being bred in zoos; curiously the leading breeding zoo is in Basel in Switzerland. The domestic form is usually called the Donkey (*Equus asinus*). The names Ass and Donkey are actually interchangeable. The word 'donkey' is etymologically obscure and only emerged in the late C18th perhaps to avoid confusion with the word 'arse'! One great importance in our dealing with them is that, unlike horses, they are not herd animals.

Donkeys were probably originally domesticated by the cattle-raising people of Nubia, and Somalia. Although their load bearing capacity is limited (20% - 30% of their body weight), the ass was found to be a more useful pack animal than the ox, which they had habitually used until then. (The ox needs a rest period in which to ruminate and is a much slower mover.) With increasing mobility an easier nomadic way of life became possible with whole families following the flocks from pasture to pasture in an overture to long distance trade. Centuries before the first pyramid was built, pack trains were journeying from the Nile Valley to the Red Sea to trade with Arabia and generally to establish commerce, as the chapter on trade routes will illustrate later.

Certainly Egypt seemed to have had domesticated donkeys as far back as 4,000BC. As proof, the skeletons of three domestic donkeys were found in an Egyptian tomb dating from that time. They also appear in wall paintings giving archaeologists evidence that the most powerful and rich Egyptians kept herds of over a thousand head. Not until later did they become the animal of the peasant farmers and that was because, by then, the elite had found something better and more upmarket – the horse.

Asses spread through all the known Middle Eastern area with Mesopotamia the main centre for breeding them. Man had, by going the tempting selective breeding route, managed to build up several different varieties. For example Damascus was well known for its large, white riding asses and in fact the nickname for Damascus was "The City of Asses". In Syria other breeds were developed, including a gentler one for women. And in the Muscat and Yemen a variety was developed which is still very much in use to this day. (Years later the white ass was the preferred mount for the Popes before the advent of motorised transport. Gifts of white asses donated to the Pope were recorded in the Harleian Miscellany, found in the Earl of Oxford's library.)

Then in 2,000BC the donkey was moved into Europe via Greece. The Greeks had exported vines to their colonies in North Africa and west into Italy so somehow the donkey became associated with the vine and therefore with Dionysus, the Syrian god of wine. They are often depicted with vine garlands on their heads. The Romans continued with the good work and took donkeys with them into France and Spain. Leaping forward in time to 1495, Christopher Columbus, on his second voyage, carried the very first donkeys into the New World. Four jacks (males) and two jennies (females) were among the inventory of livestock delivered to Hispaniola, the Caribbean Island he visited. There they were bred with horses to produce the very first mules, which played such an essential part in the Conquistador's subjugation of the Aztecs and the Incas.

The French bred their very own unique donkey the Poitou which has the distinguishing features of an exceedingly long matted coat, called a 'cadanette', a height of up to 16 hands and huge ears. The origins of this donkey are obscure but may even go back to Roman times. They were selected for size and then crossed with the Poitevin horses called "Mulassières" or Mule-breeders to produce exceptional mules. These mules remained important right up until shortly after WWII. Sicily went to the other extreme and has short-legged pygmy donkeys about 3 feet (9 hands) at the withers. They apparently are very affectionate and make lovely pets.

Ponies were brought into the British Isles by the Celts in about 2,000BC. The word 'pony' comes from their goddess Epona. There are at least eight breeds which evolved their own characteristics owing to geographical isolation. The best known ones are the Shetland, the Dartmoor, the Exmoor, the Fell, the Dales, the Welsh, the Connemara in Ireland and the New Forest. The official definition of a pony is the height at the withers which must measure less than 14.2 hands (58" or 147cm). They have stockier bodies than horses and shorter legs in proportion, and very often a heavy hairy coat as many live in extreme weather conditions. They have been put to various uses – the pony cart was a very good way of getting your children around if you lived in the country, and also provided a first step for young riders. They were often used as pack animals as they are very strong for their size. A very different existence was lived by the pit pony which was the essential forerunner to mechanisation in the coal mines of England. They replaced child and female labour. It may seem terrible to think of them living underground for months on end but they usually had a very close relationship with the man or men in charge of them and were quite disorientated by a freer life above ground. By 1913 there were roughly 70,000 employed in the mines and only as recently as the 1990's were they finally phased out.

A completely new use has recently been made of Shetlands as guide horses, taking over the same role as guide dogs.

* * * * * *

Now for the story of the true Horse (*Equus caballus*). As we have mentioned the last of them had left North America or been killed by invading man about 8,000 years before, then they gradually spread across Asia, the Middle East and into Europe. The horses which had arrived in Europe almost suffered the same extinction. There is a huge ossuary at Solutré near the town of Mâcon the remains of where Cro-Magnon man had driven horses, estimated at about 10,000 of them, over a cliff as a means of killing them. This was not on one occasion but on many, down the centuries as and when meat was required. In those times Man simply ate Horse. In the lower Palaeolithic period horse flesh was very important as food across what was to become Europe and right through to Mongolia. The horses were slaughtered in autumn when they were in good condition and the cold winter helped store the meat until the spring. The normal pattern was to kill them at about 14 or 15 years old and no longer productive of milk. The horse's fat had high calorific value and was extremely digestible even for babies. The younger horses were kept for milk. When horse's milk was fermented it was known as kumys which is mildly alcoholic. There is a Kaz (Kazakhstan) proverb "Kummys cures 40 diseases".

According to archaeozoologists by 6,000 years ago four ancestors of horses had developed, the Przwalski, the Tarpan, the Forest Horse and the Tundra horse. The results of archaeozoological investigations are vital for an understanding of human-animal relationships in the past. Somehow from this assortment developed a range of equines from ponies like the Exmoor and Shetland, to the heavy horses of northern Europe and then the faster lighter breeds like the Akhal-Teke and the Arabians.

A wild species that was not rescued in time was the Tarpan. The Tarpan (*Equus caballus ferus*) ranged from southern France and Spain up and into central Russia. In Western Europe the earliest signs of the Tarpans were found in 1995 in some dramatic and beautiful paintings in the Chauvet Caves situated in the Ardèche in France. This places them at 30,000 years ago or Upper Palaeolithic era. Other cave paintings of horses with spots were found in the Pech Merle caves near Cahors, dating back to 25,000BC. At first people were sceptical about the spots, but a recent study of horses of the last glacial times has shown that there were a huge variety of coat colours. The Lascaux Cave (17,000 years old) in south-west France, also show early horse paintings. The first description of them was by the German naturalist Johann Gmelin (1748 – 1804) who found them near Voronesh in 1769. Domesticated by the Scythians around 3,000BC, their colour was dun or smoky grey with a dark dorsal stripe that ran right through their blonde mane and tail, and they stood at about 13 hands. Their hooves were so tough they never needed shoeing. They were probably the stock from which the fabled Ferghanas were bred, and are thought to be the ancestor of the Arab and also of the Roman racing chariot horses. The Tarpans' meat was considered a great delicacy in Poland so by the

C18th they were near extinction and the last one died in the Moscow zoo in 1887. It seems extraordinary that if the meat was so prized, man would not have sent them into extinction but carefully conserved them. The Polish government belatedly created a reserve for the last of them in Bialoweiza but too late and by the late 1800's, they had died out.

Then in the 1930s, two zoologists, the Heck brothers decided to try and breed them back. They were the same brothers who had tried to resurrect the aurochs. Heinz worked at the Hellabrunn Zoo in Munich and his brother Lutz at the Berlin Zoo and their experiments began even before The 3rd Reich. This was one experiment in genetics that did not transgress the boundaries of morally acceptable science. They bred together a selection of small horse and pony breeds which they reckoned had Tarpan blood in them, the mares belonged to the Konik, the Gotland and the Icelandic breeds and the Hecks mated them with Przewalski stallions. The results do bear a very close resemblance to the Tarpans but whether you can ever 'breed back' a lost breed is a moot point.

Remains have been found in Scandinavia dating from about 10,000BC of a heavier boned and bigger horse called the Forest Horse (*Equus cabullus silvaticus*). It stood at 15 hands and had thick legs and broad feet to cope with the swampy conditions and may well have been the ancestor of the European heavy horses that developed later. But there is a good deal of controversy over these horse breeds and how they arose and their connections with each other, and whether there was more than one attempt at domestication.

Another ancient ancestor was the Tundra Horse whose remains have been found alongside those of a Mammoth and whose only known descendant is the Yakut pony. This little pony grows a thick white coat and an abundant mane in the winter which enables it to stand such extreme low temperatures (-60degF) that Captain Scott decided to use them on his ill-fated journey to the South Pole.

Of the ancient breeds of horses, we are extremely fortunate to have some representatives that escaped extinction. In 1881 a most exciting find was made by a Russian officer Colonel Nicolai Przewalski, of the last remnants of a primitive horse in the Dzungarian desert of Mongolia. Called after their finder, these Przewalski Horses or (*Equus cabullus przewalski przewalski Poliakov*) are sturdy little creatures, standing about

13 hands, of varying shades of golden brown with the identical dark, stiffly standing manes as zebras, a long heavy head and no forelock. One difference from other breeds of horses is that they have 66 chromosomes instead of the usual 64. On their legs they frequently bear vestiges of ancestral stripes. In the winter their coats turn cream for camouflage. In Mongolia the Przwalskis are known as takhi, which means 'spirit' and are preserved in Mongolia's Hustai National Park. They are also being preserved in zoos like Marwell's in Hampshire and herds of these horses are now running wild in the Cevennes and on the Hungarian plains. The USA also has a big breeding program run by The National Zoo (Smithsonian) at their research centre in Virginia. Toronto too has a breeding program thus making a diverse basis for breeding. So, rescued from the brink of extinction, about 2,500 exist today. It is not actually a direct ancestor of the Horse but it is a closely related species.

Then in 1995, a European expedition into Tibet led by a French anthropologist Michel Peissel found another previously unknown species of horse, very similar to the Przhevalski breed, which he named after the Riwoche Valley in Tibet in which it was found. Only four feet high (12 hands) and with stripes on legs and back, they closely resembled the primitive horses depicted in cave paintings. Tiny, bad-tempered but robust, they had been partially domesticated by the local Khamba tribe.

Another rather fascinating missing link story took place in Iran during the reign of the last Shah. An American woman called Louise Firiouz (married to an Iranian) was living near Persopolis in Iran. In the 1960's, she noticed a very small unusual looking horse pulling a very heavy load with no apparent effort. Intrigued, she began collecting these horses from the mountainous region near the Caspian Sea. The resemblance to the horses depicted on artefacts from the time of Darius the Great (519-486 BC) was really striking. Up until then bones found in archaeological digs had been thought to be those of onagers. Darius was so keen on these amazingly strong little horses that easily managed to pull his war chariots that he even had them depicted on his seal. Through the auspices of Firiouz the Caspian horses, as they are called, also caught the fancy of Shah Reza Pahlavi and inspired him to found a Royal Horse Society, which was visited by other royalty including Prince Philip in the 1970's. By 1974 revolution nearly put an end to it all as no one was allowed to own or breed horses. Firiouz's horses were confiscated but undaunted she built up another herd from carefully selected feral horses. Once more she ran into opposition but somehow she managed to send out enough horses to start a viable herd in England. So after 2,000 years the Caspians were saved at the brink of extinction. In 2011, the remains of a Caspian horse were found at Gohar Tappeh in Iran dating back to 3400BC. This showed the breed was even older than anyone thought. However after the Iranian Revolution of 1979 a decree was issued which only allowed one horse per man and this led to people hiding their extra stock or letting them go feral. They are still used in Iran for drawing carts but in England and the USA Caspian Horse Societies have been formed to keep the breed going.

But to return to the main theme, with the fate of *Equus* hanging in the balance about 5 or 6,000 years ago, one of the most exciting things the world has ever known happened – Man began to domesticate the real horse, for owing to its superior ability to defend himself it was the last of the *equids* to be tamed. This domestication took place on the Eurasian steppes, possibly in the Ukraine area, where people were largely nomadic and needed assistance with their movement.

Man already had dogs, sheep, goats and cows and at first horses may have been kept for milk and meat only, but then they needed pack animals and horses were the obvious choice. Pack animal is one thing but by what a leap of imagination did the first person mount and ride the first horse about 6,000 years ago? It was the most amazing act of daring for a puny man to climb on and dominate this powerful dangerous animal weighing nearly half a ton. How did it happen? Was it a great macho clash of Titans? For although those early horses were much smaller, they are never easy animals to make submit; they can kick and bite lethally; they are also very swift, and flight has always been their main strategy when threatened. Maybe it wasn't a strong and dashing youth seizing a full-blooded stallion. Perhaps a poor lame boy tamed an orphaned foal. Then one day he was inspired to seize the opportunity for a lift that carried him triumphantly past his fitter peers. That would have put him well ahead of the game. No one knows how it happened and it certainly happened more than once. Aborigine and traditional people tame and breed from hand-reared infants so starting with foals is probably likely. Excavations at a place called Dereivka on the Ukrainian steppes have yielded horse's teeth dating from

3,000BC that showed wear by some form of bit having been used. New findings in Kazakhstan by the Carnegie Museum of Natural History places domestication at least 500 years earlier. The advent of mitochondrial DNA has enabled evolutionary biologists from Uppsala University in Sweden to compare nearly 200 specimens from living primitive horses such as the Przwalski and the Icelandic with 12,000 year old bones from the Alaskan permafrost. This showed that horses had been domesticated in many places and at many times. Both the Brahmins in India and the Chinese claim to have begun riding in 4,000BC and there is strong evidence of the Assyrians using horses, and the first actual pictorial record of riding was found in Persia dating from 3,000BC. By 1580BC horses appeared in Egypt and 250 years later in Greece.

Anyway once riding started, it caught on, spread extremely quickly and soon it transformed human society and of course warfare. From then on horses were up and away and we with them, and a bond that was to last 5,000 years had been sealed. They now were truly valued, for far more than just milk and meat. Back they spread through their old domains, from the Ukraine back to sites from where they had long vanished, west through the Caucasus, Hungary, and Romania and finally they returned to Europe.

But there was so much to be done before those wild, galloping, kicking creatures could become the horses we know today. It was the equivalent of turning the wolf into the assorted breeds of Labrador, terrier and poodle. Horse-breeding became absolutely vital. Everyone wanted to upgrade and improve the breeding stock of the original small stocky ponies. Warfare, trade, and hunting – everything now began to depend on the horse. You can imagine they must have become the most dominant topic of planning, of conversation, of dreams.

There were many reasons for selective breeding. There was no doubt that as always, snob appeal was a motive all on its own. But keeping up with the Joneses was absolutely imperative especially when the Joneses wanted to make war on you. As has been seen, various breeds of horses were developing in different parts of Asia and Europe. So those countries that lacked suitable animals started importing horses from other areas to improve their home grown stock, sometimes having to use force. Selective breeding was starting up in numerous ways, of various kinds and at different times all over Asia, China, the Middle East and then Europe. And the great distinction from any other selective breeding of animals is that horses were the only animals to be bred to be larger. Dogs, pigs, cattle, all became smaller as well as more tractable. But the horse had to carry a man or pull a cart and a tiny Przewalski horse would not have coped with the roles that man was selecting for it. One can only mention a few examples.

A broad classification grew up dividing horses into 'hot-blooded' and 'cold-blooded', based on their use and their temperament. Into the 'hot-blooded' category fell the more spirited horses that would eventually lead to thoroughbreds and racing stock, as well as the preferred steed for cavalry – the Arabs, the Ferghanas, the Akhal-Teke and others. The 'cold-blooded' horses denoted the heavy draught horses for wagon and plough and for war.

The Akhal-Teke of Turkoman stock came from the area of the Parthian capital Nissa (Turkmenistan). In the Akhal Oasis near Nissa, horse shaped jugs have been found and Terracotta statues of chariots dating back to almost 2,500 years ago. They were the horses of the Scythians who had great cavalries as long ago as 700BC and who began selective breeding.

Some of the best of the so-called hot-blooded horses were the fabled Ferghana Horses, from a valley north of the Pamirs near Uzbekistan, in Kazakhstan. Just how did beautiful, fast long-legged creatures evolve from the stocky little ponies? It is another equine breeding mystery. But they were so prized by the Chinese, that they were symbolically called "The Celestial Horses". A feature of their particular build was their front legs being longer than their back legs. They were rumoured to produce red sweat like blood (possibly a skin parasite). Remains of their skeletons have been found in Pasryk, buried in the graves of ancient warlords dating as far back as 3,000BC. Naturally these strangely beautiful heavenly horses became sought after all over Asia. By 1600BC the fast chariot was in full use for warfare and this time is known as 'the chariot age' and lasted about a thousand years. Full armour for both man and horse was being used and the weapons varied from bows and arrows, to javelins and spears. Most of this was brought about by the Hurrian people of Syria and Mesopotamia.

By about 500BC the Chinese were getting desperate concerning the repeated raiding by the Hsiung-nu (later called Huns) on their fast little Hunnish, Mongolian type ponies. The reaction of the Chinese was amazing, as a defence they began building the Great Wall of China. The first 1,500 miles of it was completed in 209BC, but still the raids continued. The 'Huns' practically lived on horseback but the Chinese were indifferent riders and their horses were ill-bred. They were no match for the invaders. Then the Emperor Wu-Ti (145-87BC) decided that the only possible way to beat the nomad invaders at their own game was by upgrading the Chinese pony. Wu-Ti was also worried about his fabled immortality which required him to be drawn up to heaven by the heavenly horses and about which he was beginning to have doubts. So he decided that the solution to both these problems was to acquire some of the "celestial" Ferghana Horses. But there were two major problems – Ferghana was 3,000 miles away and the King of Ferghana refused to sell any of his horses. Wu-Ti dispatched an ambassador to barter a golden horse in exchange but he was killed right at the frontier. The second expedition consisted of an army sent over the Pamirs but the great distance and lack of fodder (locusts had ravaged the grasslands) reduced them greatly. Finally, the third time and with nothing left to chance, Wu-Ti laid siege to Tashkent. The inhabitants gave in, killed their king and let the Chinese choose a thousand breeding Ferghanas. This also solved his personal problem – Wu-Ti now expected to be drawn up to Heaven by the heavenly horses and given one of the peaches that ripen only once in every 6,000 years, and this would bestow immortality upon him.

The Akhal-Teke, the Ferghanas and the Arabs may have been just variations on the same theme – the fast spirited horse, bred in wide open spaces which used its speed as an essential device to escape trouble. Philip II of Macedon (382-336BC), another famous conqueror, also imported twenty thousand of the Ferghana mares to improve the Greek bloodstock. And yet another was the Emperor Zu Dhi in the C15th, who by buying so many thousands of these horses put such a strain on the treasury that a special 'Tea For Horses' bureau was established, using tea in place of silver to pay for his extravagance. The 'Tea and Horse' route became one of the great trade routes second only to the Silk and Spice Routes. Even today there are estimated to be about 11 million horses of at least twenty-six breeds in China, which makes up about 1/6th of the world total.

The Arab Horse was the other related breed that was and is highly valued right down to today. It seems the Arabs prized their horses far beyond their wives! According to the 19th century veteran Arabian traveller

Lady Anne Blunt, when writing of the people of the Nejd desert: 'Truly is that horse prized by them above all else in the world' You may hear [them] remark: 'Children of mine may hunger and thirst, but never my mare.'

Their origin also is obscure, but they probably existed on the Arabian Peninsula as far back as 3,500BC, and such is the aridity of the desert that they must have co-existed with Man. The Arab horse, like its master was hot-blooded. The Arabs gushed out poetry over horses. "The horse was the wind and her master put a bridle on it". An unknown Arab poet of long ago wrote:-

"The nostrils of a racer are like petals of a rose. . . . The neck is an elongated wave from which floats brilliant ripples of silken mane. . . . The ears, inward pointing, are lilies in trembling water, and the whole body of the mythical, yet fleshly horse sways with the supple strength of wind, sun, and sand."

"A drinker of the wind, a dancer of fire," is another Arabic quotation. The early Bedouin lived a life consisting entirely of fighting and raiding. Horses were the prize booty. The pure bred were called *asil* and the mixed breed was *kadish* and of no importance. They were obsessed with purity of line and believed that if a mare gave birth to a foal from a *kadish* stallion, then any further mating with an *asil* stallion would be tainted. In fact so worried about their mares were they that before raiding enemy territory they would sew up their mares' vaginas to protect them from impure mating. Oral records were passed down from father to son until about 1330AD when written records began to be kept that used the title 'Arabian'. Stallions were the choice for warfare as their natural aggressiveness could be directed against the enemy.

Horses figured greatly in the Bible and the Qu'ran. According to biblical legend breeders of horses started with Adam, then Ishmael, the son of Abraham, Baz the great-great grandson of Noah, Solomon and finally the Prophet Muhammad. The last two really did advance the cause of the horse. King Solomon (970 – 931BC) ignored the Israelite law forbidding the keeping of horses which, not surprisingly, was considered to be a form of idolatry. He went to the other extreme with a stable of 40,000 horses of Arabian stock, 4,000 chariots and 12,000 horsemen. So at his command he had a formidable mobile army. He also ignored Moses' curious decree that the king should not breed horses. Muhammad turned farmers and stock breeders into warriors. His own horse Al Borak which meant Lightening was rumoured to possess a human face and was to bear him up to heaven, when he died.

In 1,200BC the mysterious 'Sea Peoples' arrived in North Africa and domesticated the ancestors of the famed 'Barb' or Berber breed. Hannibal used these when he fought against Rome and a thousand years after his death the Moors were to use the same type of horse when they invaded Spain, but curiously they preferred to use mares.

Alexander the Great's famous horse Bucephallus was supposed to have been of Akhal-Teke stock, and with this mount he conquered Samarkand, which is actually part of the Ferghana valley. From the C5thBC on, the Greeks and the Arabs of central Asia valued these animals and called them 'Elite Horses' and 'The Greyhounds of the North'.

Then the Greeks also became keen on importing the swift Arabian horses and with these spirited horses incorporated Chariot racing into their Olympic games. As the status of the horse rose it became more and more associated with the nobility, with kings and even with gods. The social divide was obvious for all to see, the nobility were equestrians and all the rest were pedestrians. In classical Greece your income group and status were indicated by words such as 'hippeus' which meant a man wealthy enough to own a horse as opposed to 'zeugites' which indicated a man plus a pair of oxen.

The Romans, preferring their legions, may have used mercenaries for their cavalry but they did start a policy of selective breeding for all their special needs – harness, hunting, racing – especially drawing chariots. But they failed to take up the innovations of saddles and stirrups which no doubt helped Attila the Hun with his superior cavalry annihilate them in the last battle of the Empire in 451AD.

When Charles Martel, the French King, drove back the Muslim invaders in central France at the Battle of Poitiers in the year 732AD he managed to capture a number of Arab horses from his defeated enemy to improve his breeding stock. He wisely took a great interest in horse breeding. Because for warfare, for power, for demonstrating status the horse had no equal; in his different guises he was the Rolls Royce, the Lamborghini, and the 4 by 4 of the times. And in warfare he was the equivalent of airpower unleashed on unarmed civilians.

The Bedouin practiced a kind of accidental form of selective breeding, for when they raced their best horses the prize for the winners was the pick of the finest stock from the loser's herd. This led to improved diversity. Arabian horses, like the Chukchi dogs of Siberia were part of the family and they too slept in their master's tents and even ate from the same bowls. After a foal was weaned from its mother, it would be fed on camel's milk for 100 days and then it was gradually introduced to barley. A horse can live only three days without water so if no water or pasture was available then the Bedouin would feed them camel's milk and dates. In an early form of racing, the horses were deliberately kept thirsty and then they took part in a race to the nearest water.

But all in all the nobility owned the horse and the horse owned the nobility. No wonder 'egalitarian' revolutions such as the French, the Russian and the Iranian all targeted the horse as representing privilege. All of these repressive regimes made it illegal to own or breed horses. But the consequences could be dire. For example the French Revolution's prohibition of horse-breeding meant that the English had heavier and better horses, which must have contributed greatly to the victory at Waterloo. Another example of this treatment took place in Poland where an Arabian stud farm, called Janow Podlaski was established in 1817 but it too was interrupted and all but annihilated by first the Russians in 1914 and later by the Germans in 1939. Luckily, enough breeding stock survived and the stud has been re-established since 1960 and is going strong today. Count Potocki was less fortunate. He died in 1918 defending his stud from the Bolsheviks.

Arabian horses were carefully bred during the last Shah's reign and are still being bred in Iran. They were also very popular in Russia and there were stud farms started by Peter the Great whose favourite mare Lisett was of Arabian stock.

In the late C19[th] an unexpected rescue of the Arabian horse breed was launched by Lady Anne (granddaughter of Lord Byron) and Wilfrid Scawen Blunt. In 1879 they crossed the great Nafud desert in the heart of Arabia. They continued to travel all over Arabia and the Middle East, buying Arabian horse from the Bedouin. They then started a stud for purebred Arabs in Sussex at their home Crabbet Park and largely thanks to them the purity of the Arabian breed was maintained. One of the most famous stallions was Mesaoud originally owned by an Egyptian Ali Pasha Sherif. The lineage of most of today's Arabian horses can be traced back to Crabbet Park and mainly the care of Lady Anne who was far better than her husband at caring for the Arab stud.

Apart from these fast dashing horses other breeds were evolving – strains that were more heavily built, slower and stronger – the ancestors of the Percherons and Clydesdales as well as the Suffolk Punches. These had been bred heavy-duty enough to carry the mediaeval knights in their heavy armour. Then about 1,000 years later the Crusaders returned with more Arab horses to really breed the Percheron as we know it. Henry VIII actually made laws to encourage the breeding of larger horses by importing selected stock from the continent. The Moors also brought their own horses to Spain, when they conquered it. With crossbreeding, a magnificent white strain emerged but by their laws no one but a Moor could own one of the white horses. It was nearly lost again in the equine bloodletting of the French Revolution but survived with the name 'Percheron'. They were imported into America and South Africa in the 1840's and became the favourite draught horse in both countries and made a great contribution to development. By 1915 over 40,000 Percherons were registered in the USA.

During mediaeval times horses in England were given names that referred more to their use rather than the breed. Probably the best known is the Destrier also called the Great Horse which was used as a charger in war and as a hunter in peacetime. It was extremely expensive compared to ordinary horses and mainly used by the nobles. Then there was the Courser, light and very strong and also good in battle. The squires and men-at-arms were assigned the Rouncey, which was sometimes used as a pack horse too. Another upper-class horse was the Palfrey, used very much by the Ladies. One of the great things about it was that it could be induced to adopt the ambling gait, so much more comfortable than trotting for the rider on a long journey. Another type was the Hobby, light weight and only 13 to 14 hands but very agile and good for skirmishes so used in a specialised cavalry called Hobelaars. The Sumpter was simply a packhorse and the Spanish Jennet was another type of riding horse with a smooth gait.

The Clydesdale at a towering 18 hands (6ft) and weighing 2,000 lbs makes a striking sight in the English countryside with its distinctive feathery hairy fetlocks. For its size it is amazingly graceful and very gentle. The story goes that Queen Elizabeth II, the great horse lover, saw one pulling a milk cart and was so struck by its presence that she had them introduced as drum carriers for the Household Cavalry.

Another slightly smaller heavy horse is the Suffolk Punch which is around 15 hands and is always chestnut. It is one of the oldest English breeds and was used from the early C16th for farm work and pulling artillery. A Suffolk Horse Society was started in 1877 and soon afterwards, Suffolk Punches were imported into North America. With the advent of tractors they were on the point of dying out and in spite of a revival are still on the critical list at the 'Rare Breeds Survival Trust'. In the early C20th some were exported to Pakistan to improve the breeding stock there.

The Cleveland Bay is supposedly the oldest breed of horse in England and was originally an all-purpose horse extensively used by the travelling salesmen called 'Chapmen' in mediaeval times to carry their goods. The name Chapman is a cognate of the German Kaufman. So at first they were known as Chapman horses but later took on the name of Cleveland Bay from a combination of their home area and colouration. They also added the job of coach horses to ploughing, hunting and pulling carts. In fact they are the perfect example of an animal that helped Man with Movement in every way.

Then we see the finest example of equine engineering. Cleveland Bay mares were combined with imported Arabian stallions to create the English Thoroughbred. The term "Thoroughbred" referred to horses in the General Stud Book. The three iconic stallions were the Byerly Turk, the Darley Arabian and the Godolphin Barb.

These three stallions all have extraordinary stories behind their introduction to the making of the thoroughbred. Following their family trees is rather like compiling a 'Who's Who' of racing champions! In 1686 Captain Robert Byerley captured a stallion of mostly Middle Eastern blood at the Battle of Buda from the Ottoman enemy. He rode this stallion through the remainder of the war and also in Ireland at the Battle of the Boyne. Luckily the stallion survived and was taken to Goldsborough Hall in N. Yorkshire where he began to breed it with the native mares. The horse lived for six more years siring many important horses and becoming the greatest contributor to the Herod line of thoroughbreds along with

Highflyer the son of Herod. This stallion is entered in the studbooks as Byerly Turk, the last 'e' being accidentally omitted.

The second stallion is the Godolphin Arabian or Godolphin Barb (Barbary was the old name for Western North Africa) and was foaled about 1724 and probably exported from Yemen via Syria to the stud of the Bey of Tunis. Initially given to Louis XV of France in 1730, he was then brought to England in 1728 by Lord Godolphin and was the sire of Lath, the best racehorse of the day. He was quite small but beautifully built and his line goes down to Eclipse the famous fastest horse of his day. George Stubbs painted a beautiful portrait of him.

The third stallion known as The Darley Arabian had a most adventurous start to his life. He started life in the Syrian Desert near Aleppo and belonged to Sheik Mirza II. The British Consul, Thomas Darley tried to buy him, but The Sheik reneged on the deal. The stallion was then smuggled out by British sailors via Smyrna and shipped to England. He was taken to the Darley family seat Aldby Hall in North Yorkshire where he was put to stud and showed great stamina from 1706-1719. This stallion was selectively mated to some of the native racing mares and the first great racing horse was born – Flying Childers. His line also runs through Eclipse.

Another isolated group are the Horses of the Camargue which are believed to be descended from prehistoric herds crossbred with Berbers from North Africa and used by the Romans 2,000 years ago. Over the centuries they have evolved large feet with flat hooves to cope with the marshy ground. Apart from the rough grasses their diet includes samphire, a plant that grows on salty marshes. They became the traditional mount of the farm workers or ranch hands, who rounded up the wild black bulls for counting and branding. In 1967 they became recognised as a breed in their own right.

But not everyone was after speedy, dashing horses; small tough ponies were and still are in demand. Mongolia alone has over 3 million of them, probably related to the extinct Tarpans. They actually outnumber humans and their devoted Mongolian owners apparently use over five hundred terms to describe their horse coats. These incredibly hardy Tartar horses can dig through the snow to find grass. Sadly they were nearly exterminated, by order of the Soviet authorities, who wanted to replace them with machines and return to the earlier habit of eating horseflesh. The central Asian ponies were very like their forebears the Przewalski horses, tough, ugly but easily domesticated. It was reported by more than one observer including Sven Hedin, the famous explorer that the Tibetans fed their horses with meat. They also had a very curious way of catching their horses. Instead of using a lasso they would creep up to them then place a hand on the right front leg, slowly moving it up and probably caressing the horse, until they could grasp the bridle.

The Persians were the first to develop communication using horses in the earliest form of "pony express" by placing posting stations one day's ride apart. Messages could be sent from Susa to Sardis across the Persian Empire, a journey of 1,600 miles in nine days. This messenger service of the C5th BC was not bettered until Napoleon's day.

In the rest of Europe and England, horses were still essential for everyday activities like draught, carriages, hunting, racing, and general transport. In France Louis XIV established stud farms and bought breeding stock from other European countries and the Barbary Coast. By 1789 it is estimated there were 21,000 horses in Paris which led to the most appalling congestion not to mention fouling of the roads.

One result of all this breeding is that at the present time there are horses varying in size from the smallest, a Falabella, just under 5 hands (19 inches) weighing 30lb to the tallest, a Belgian draught horse of 19.2 hands (6ft 6ins.) that weighed 3,200lbs. All over the world now there are numerous breeds of horses – about 300; some are being developed, others are disappearing. According to the 2007's census there are 58 million horses in the world. Ten countries have more than a million.

Another great innovation by man was the mule. Man brought together horses from Eurasia and donkeys originally from Africa and somehow the idea came to cross breed them, or maybe they had the idea themselves! Horses and donkeys certainly co-existed on the Nubian borders around 1,750BC. There has been cross breeding the onager with the Ass and even the Donkey with the zebra to produce zebroids but it was the mule that was to prove so useful down the ages. The proper mule is a cross between a horse mare and a donkey sire or jackass. Hinnies are the opposite – a stallion horse crossed with a donkey jennet. Being purely man-made animals, with few exceptions, they cannot reproduce. The curious thing is that a mare will produce a bigger foal if sired by a donkey than if sired by a horse, a real demonstration of hybrid vigour.

Models of mules appeared in the pyramids and on Assyrian monuments. The Bible contains many references to mules and Jews were forbidden by their religion to breed them but curiously they were allowed to use them. The Hittites, who were so keen on horses, actually prized mules at three times the price of a horse – 20 sheep for a horse and 60 sheep for a mule. This was partly because their wild Tarpan horses were only thirteen hands and the mules grew at least a hand taller and were useful as a beast of burden and for ploughing as well as riding. According to figures brought out by the Animal Transportation Officer in 1988, they are stronger than horses and can carry roughly one third their body weight so that could be as much as 200-270lbs. Their skin is harder and tougher than a horse's and so less inclined to chafe, their hooves are also stronger. Another of their strong points is their surefootedness over the difficult terrains of such disparate places as the Himalayas and the South African Veld. One typical perversity is that they go up hill faster than down. Mules exhibit the best characteristics of both their parents, the strength and courage of the horse combined with the stamina and light-footedness of the donkey. They are also more intelligent and sensibly avoid unnecessary risk. Alexander the Great must have thought highly of them as he had a chariot drawn by twelve mules. In Mediaeval Europe horses may have carried the armoured knights, but mules were the choice of gentlemen and clergy. Jesus may have ridden his donkey into Jerusalem but Mohammed is recorded as having ridden a mule into battle. Popes always used to ride mules.

Mules and oxen were the main sources of traction power. They were the preferred carriage animals for pulling queens and aristocratic ladies in their two-wheeled covered wagons or *carpentum*. They were also the mainstay of the Roman postal and public transport services.

Another use was found for them – pulling barges, a precedent for their use with barges in England from the C18[th] until the mid-C20[th]. Marco Polo (1274) lavished praise on the Turkoman mules of Central Asia, and they certainly existed in Britain in the C12[th] – one actually appears on the Bayeux Tapestry. By the C19[th] the English, exhibiting a strange prejudice had gone off mules almost entirely. Yet for ordinary every day purposes they had more endurance, were cheaper to keep, and were stronger, less disease-prone and lived longer. In Ireland jennets were very popular.

The Spaniards have always been particularly keen on mules. Columbus took four jacks (males) and two jennies (females) to the New World in 1495. More were brought over in time for the conquistadors to conquer the Aztecs and the Incas. Males were used as pack animals and the females were preferred as riding animals. They were much stronger than the native llamas and could cross sandy coastal areas that the llamas could not manage. By 1776, some 500,000 mules were employed in Peru for trading along the coast and for drawing carriages in Lima. In fact in South America by the end of the C18th there must have been close on two million mules mostly for riding and as pack animals. Horses and oxen were two a penny and a mule was far more valuable.

They were not much used in North America until George Washington encouraged their use and was sent some good breeding stock by his old friend the Marquis de Lafayette. The King of Spain sent him a large Spanish jack named 'Royal Gift' that was used to cover mares and jennies. By 1889 there were an estimated 157,000 in America. "Forty acres and a mule" was all that was needed to make you self sufficient. Their endurance and strength was amply demonstrated during the Civil War. Other imports of jacks from Spain led to their own race of Jacks called 'The Kentucky Jack' breeding mules up to sixteen hands. From then on they were of immense value in the opening up of the American West. In the Pennsylvanian mountainous areas they worked in terrain no horse could have managed. They were also used to draw the 'Prairie Schooners' as the covered wagons were called. After the Civil War the mule continued as the major draught animal until the advent of the petrol driven tractor. Even Buffalo Bill Cody had a little Comanche mule called Mouse, which impressed Custer so much that he tried to swap his horse for Cody's mule – but in vain.

There has been an annual horse fair at Rawalpindi in India since 1876 with up to 1400 mules for sale. At first, out of superstition, the Indians refused to breed mules. But when they found that a worthless mare produced worthless foals but acceptable mules they changed their breeding program – a classic case of money influencing religion.

But mules do have their quirks. If brought up by a mare they tend to hate asses. A horse, preferably a white mare acting as half the lead pair pulling a wagon, will greatly improve a mule's performance. They seem to have 'a thing' about white mares as one with a bell round her neck will keep mules from straying when they are out spanned. Another quirk is they can be nervous of strangers.

Horses, mules and donkeys all had their role to play in our lives for nearly 6,000 years. Mechanization has of course replaced the majority in the western world but they still have their essential roles to play in the third world. And even in the west the horse is one of our main playthings for hunting, racing, games and processions. A beautiful horse is still a thrilling sight even to the uninitiated.

The equines had moved out of our larder and entered our short list of beasts of burden but they had to be trained to do our bidding, to pull wagons and coaches, to drag ploughs and most importantly to be ridden in a multitude of situations. This is dealt with in the next chapter.

ET CETERA

1. In Ireland, in a form of snobbery or racism, horses were given English names and cows were given Irish ones.
2. Cosimo de Medici nearly ruined himself reinstalling a guard of 2,000 horses in 1531.
3. The famous White Horse cut out of the chalk downs at Uffington is 374 feet long and 3,000 years old.
4. <u>Hobson's Choice</u>. Thomas Hobson was a humble stable manager in the C16th who placed his most rested horses nearest the stable door and gave his customers no choice. He made so much money he ended up buying Anglesey Priory.
5. <u>Horse Power</u> is still used to compare the output of a steam engine with the power of draught horses. One horsepower = 746 watts. (James Watt (1736-1819) was the inventor of the steam engine)
6. Horseshoes nailed in U shape mean good luck cannot run out. Nelson had one nailed to HMS Victory's mast.
7. Our language too pays homage with numerous expressions and similes such as Horse Sense, Horse of a different Colour, Mare's Nest, Nightmare, The Four Horsemen of the Apocalypse, As strong as a horse, Flogging a dead horse, Hold your horses! Don't

look a gift horse in the mouth. Straight from the horse's mouth, To back the wrong horse, To go by Shanks Mare, You can lead a horse to water but you can't make him drink, To ride a High Horse, Straight from the Horse's Mouth and so many others.

8. Caligula the Roman Emperor (AD12 – AD41) owned a horse called Incitatus, which means 'spurred on'. He made his horse a priest and a consul and gave him a manger made of ivory and a water pail of gold.

9. During the reign of the Roman emperor, Nero (AD37 – 68AD) the Roman racing stables went on strike and the strike was only broken by the threat to replace the horses with dogs.

10. Dionysus is pictured riding a mule which being a sterile animal seems a curious choice for the god of fertility!

11. Quote from the Quran. "Allah has given you horses, mules and donkeys, which you may ride or use as ornaments…splendid horses are amongst the comforts of this life."

12. Feral horses have their special nicknames such as the Mustangs in the USA and the Brumbies of Australia, the Sorraia and Garrano of Portugal.

13. A similar man-made event to the ossuary of Salutré happened for a different reason when during the Gold Rush of the C19[th] thousands of horses were driven to their deaths over the cliffs of Santa Barbara in California to make way for the expanding herds of cattle.

---ooO0Ooo---

CHAPTER 4

TACK, WHEELS AND TRAINING

Tack, Wheels and Training...BRIDLES. SADDLES

*"The armoured Persian horsemen and their death
dealing chariots were invincible. No man dared face them."*
Herodotus (484 - 430 BC)

Now for one of the greatest innovations of all time – the Wheel! By chance, it was invented at about the same time as the horse was being domesticated, in about 4,000BC. In fact there is a certain amount of controversy about which came first – the horse being ridden or the horse being driven. At Dereivka, a Ukrainian settlement dating about 4,500 BC, some useful remains were found – the skeleton of a stallion with definite signs of wear by a bit on his teeth. More examples are continually turning up.

So, by putting the Horse and the Wheel together, civilisation literally drove off and progress simply accelerated. The domestication of the Horse for riding and as a chariot animal is one of the most defining events in the history of mankind. Its importance cannot be over emphasized. However, it wasn't until c.1700 B.C. that man thought to pair the horse with the chariot, this being the idea of the Hyksos peoples of Egypt. The Hyksos were Semitic, and their rise to power in Egypt resulted in the 15-16th dynasties c. 1000 BC. Their chariots were made out of bronze which must have been extremely heavy. The horses were trained specifically to draw these chariots. This combination of horse and chariot made the Hyksos dynasty the most powerful of the time.

Man had already invented the plough – the most vital innovation in agriculture. But it was the wheel plus the horse that gave man, for the first time a means of rapid locomotion and without Man being the load bearer. No longer was man stuck in rural life, or just drifting along in the nomadic life; he was out and hitting the road.

Before the wheel, goods were dragged about by slide cars which limited the bearing surface to a pair of narrow horizontal runners. The slide car is tilted from point of harnessing to the ground so that only the points are dragging the weight – not unlike the Canadian travois. Then early men thought of placing runners under heavy loads rather like primitive sledges. The next innovation was combining the sledge and the rollers.

It is a mystery just when and who actually invented the wheel. It may even have started as a Potter's Wheel. But the oldest Wheel for transport, found in archaeological excavations, dates back 5,500 years ago in Mesopotamia, so it was probably the Sumerians who came up with the idea. They had a tripartite wheel composed of an oval piece enclosed by two tapering outer banana shaped pieces, pegged together. This method meant three quite small planks could make a good-sized wheel. Other solid wheels were found in S. Russia and the Ukraine. They started as rollers placed under a heavy object. The Wheel also reached ancient India. China was another country that had the wheel and may even have invented it independently, probably in about 2,800BC. Nonetheless it was never invented in the Americas, Australia or sub-Saharan Africa, nor did it reach them until the Europeans brought it with them in the C17[th] and C18[th].

The wheel had other uses apart from carts and chariots. It was used to draw water, make pottery and grind wheat. It also became the central metaphor in Buddhism that maintains that life is an endless cycle of births and rebirths.

The next important innovation was to move from the solid wheel. Mobility was increased out of all recognition by 3,500BC, when wheels were fixed on carts and chariots. After cutting away the wood in the middle to form an axle, the following significant advance came when the Assyrians and the Egyptians made wheels with spokes in about 2,000BC. The joinery had now become very intricate and required imported woods such as elm from Syria. The spokes were created in two halves and half of each spoke was an integral part of the hub. In India wheels dating from about 1,500BC have been discovered. By 1,300BC, the Chinese had also perfected the use of the chariot. So from now on, chariots and carts were not just for warfare and for moving large numbers of men, mainly warriors but also for transport of goods. They became a major instrument in terms of hunting, fighting and social structures. The differences between the chariot and the wagon were size and speed. The smaller lighter horse drawn chariot could go at 20mph, ten times the speed of oxen which drew a wagon at 2mph. And of course, much later the Romans produced a fantastic array of wheels for carriages and carts, for war chariots, quadrigas and passenger coaches.

Before the horse, the ox and the onager had been the main draught animals. Then gradually, because of its speed and endurance the horse became the favourite. And here Man made the most amazing error, and one that persisted for thousands of years. Having decided the horse would make an excellent draught animal as well as for riding, Man simply moved the ox yoke onto the horse. But the ox yoke, owing to the difference in the ox's physiology to the horse, proved to be most unsuitable. This early throat and girth harness consisted of a girth circling the belly with the point of traction being at the top. To prevent the girth slipping backwards a throat strap was added, crossing the withers diagonally. This throat-girth harness compressed the trachea with the result that it suffocated the horse as soon as it pulled against its harness. Yet the most extraordinary thing about this inappropriate harness was how long it continued to be used. It first appears 5,000 years ago in ancient Chaldean pictures, in Assyria, then in Egypt from at least 1,500 BC throughout the Minoan, Greek and Roman times. It even persisted in parts of Western Europe right up to 600AD. This was one of the most glaring examples of the innovative powers of Man deserting him. So the invention of the Horse Collar was in its way almost as important as the later invention of the stirrup. A team in modern harness can pull 4/5 times the weight as that shifted by the restrictive ancient yoke harness. It was actually the Chinese who made one of the great breakthroughs in equestrian history and invented a more appropriate breast strap harness. A theory exists that this harnessing technique was derived from the ancient human method for hauling boats on Chinese rivers and canals. Perhaps the haulers realised that what was comfortable for them would also work for the horse. Strangely the Scandinavians were very early employers of horses as farm animals and examples of metal horse collars have been found in Swedish tombs of the C9[th]AD. And so the yoke gradually fell away.

The next development was around the Classical Age (600 – 200AD) when the central pole gave way to the curving S shaped shafts found on the Han dynasty chariots. The Chinese also invented the breeching strap which acts as a sort of brake on a load. This strap is round the horse's bottom and tightens, acting as a brake when the cart tries to run forward when going downhill.

Once the horse had moved on from early domestication and been developed into faster and taller breeds, it had to be controlled whether they were pulling chariots, for draught or being ridden. The next essential was to train them to do what man wanted. Not only was training essential but all the trappings of riding that we take for granted had to be invented. Man started creating ways of making riding easier, safer and stronger – harness and tack, bridles and bits, saddles and stirrups and horseshoes. Of these innovations, bridles came first and the earliest form of those was probably a halter of woven grass round the nose. A headstall on a horse's head gives much more control than a collar round the neck. The Mesopotamians were using a nose-ring in 2,300BC probably because they had already found them effective on oxen and onagers. But for the real control needed when riding the most essential device was the bit. As already seen the horse teeth found at Dereivka showed wear from rope bits and this discovery suggests there may even have been soft bits used by nomadic tribes of the area as far back as 4,000 BC, with soft mouth pieces and cheek pieces made from bone or antler. Meanwhile the Egyptians perfected reins, bridles and bits for their warhorses pulling chariots. Funnily enough the Numidian cavalry had no bridles! They guided their horses by a stick tapped on their neck.

(To this day donkeys in Tunisia have no halter and are just guided by a tap on the rump.)

The physiognomy of a horse's head happens to be most suitably constructed as there is a most convenient gap, called a diastema, between the incisors and the pre-molars almost as though designed for taking a bit. After 2,000BC bridle fittings made of bone and antler were supplemented by metal pieces. Around 1,500BC, two types of metal snaffles appeared at roughly the same time. A snaffle acts through pressure on the corners of the horse's mouth. Bits dating from the Bronze Age have been found in various caves in Europe. And early metal ones have been found in Celtic sites from 300BC. The first bridles fashioned from rope sinew or rawhide with the cheek-pieces made of bone or antler were found in the Near East around 1,500BC. The curb bit was used in the Classical period and was dropped until the C14[th] and then returned when the mediaeval knights used two reins – very similar to the modern double bridle.

Martingales are leather straps used to control the head carriage and prevent the horse flinging its head back and injuring the rider. It has two variations the standing and the running. The Standing Martingale is very simply a strap from the back of the nose band running between the horse's front legs and attached to the girth. It is fixed and cannot be loosened in an emergency. On the other hand the Running Martingale is attached to the girth, comes through the horse's front legs but then divides into two straps ending in rings through which the rein passes and holds them in a straight line. This is far safer as the rider can slacken the reins in an emergency and loosen the pull of the straps.

An interesting variation on the bridle is the Hackamore which is simply a bridle without a bit. Control of the horse by pressure on the nose and chin. It is used if the horse is suffering from a sore mouth or damaged teeth. The special noseband is sometimes called a bosal and it can be used quite harshly as a horse's face is very sensitive.

Now for the invention of the saddle: riding bareback is both uncomfortable and unsafe. At first blankets or saddlecloths were strapped on. Who made the first saddles? The Assyrians may have been the first, or the Scythians who had numnahs of felt and leather forming two cushions fitting on either side of the spine and connected in the middle. This had the desired effect of keeping the weight off and protecting the spine of the horse; it was the forerunner of the saddle. Another great riding people were the Sarmatians, a nomadic group of Central Asia. Herodotus gave out the pleasing myth that they were descended from Scythian fathers and Amazon mothers, not surprisingly as they were famous for their riding prowess, the women as well as the men. In about the C4th BC the Sarmartians produced the actual forerunner of today's saddle on a wooden frame tipping up at the back in a cantle to prevent the rider being knocked off backwards. Yet strangely enough, the Greeks in 500BC although they had mounted archers, still had no saddle or stirrups, as can be seen galloping along the Parthenon frieze. But luckily they, even without stirrups, did not lose a battle, and that saved the West. Constant fighting with the Huns caused the Chinese to emulate the riding habits of their persistent enemy and adopt the saddle and also improve their breeding stock and the size of their horses.

The packsaddles used by Himalayan caravaneers were of several kinds. Wooden saddles were used in Ladakh and Tibet, but the Turkestani packsaddle was much superior. This was made from two thick saddle-clothes, on which was fixed a saddle cushion. The cushion was u-shaped and made of sheepskin or cloth stuffed with straw; the load was fixed to the saddle by ropes. Camel wool ropes were better than hemp or even sheep's wool.

A mysterious Nubian race who fought against The Romans from C3rd AD were known as the X Group or the Blemmyes. They had the most sophisticated saddles with pommels in front, decorated with silver and cantles in the backs and their cavalry, pre-stirrup, was amazingly effective. These saddles are very similar to those used by Arabs today. Their other more legendary reputation was that they were a tribe of acephalous monsters which means they had no heads but had their eyes and mouth in the middle of their chest.

Cushioned saddles have been found in Scythian tombs from as far back as C5th BC, and one was decorated with animal motifs in leather and gold.

Finally down the centuries there developed two sorts of saddle for riders – the Moorish, which was very similar to that used by cowboys and gauchos and the second one the Hungarian, which was the forerunner of the European solid-treed saddle, which distributes the weight of the rider over a wider area and is far more comfortable for the horse. The English saddle is now used all over the world and not just in England; there are different variations for eventing, polo and dressage. The Western or Stock saddle is the one connected with cowboys and has a wide spread. It is often very ornamental and of course has a horn on the front to attach the lasso. Asian saddles have developed from the original Scythian and Sarmatian saddles and have high pommel and cantle (the upward sweep at the back of the saddle) which are to keep the rider in place during warfare. The Japanese saddle has very ornate decorations in lacquer and mother of pearl or gold leaf and were originally used by the samurai for fighting on horseback.

A form of transport that preceded the side saddle for women was the Whirlicote, also used by Richard II, which was a kind of early coach. Then there were litters carried between horses or mules 3' wide and poles 16' long, acting as shafts supported by saddles or backstraps. French used a 'cacolet' which was like a pair of arm chairs on either side of the saddle and was especially good for carrying wounded.

The next development in the saddle line was strangely motivated by women's modesty as perceived in the Middle Ages. Women had happily ridden astride up until it strangely became improper for a lady to straddle a horse. The first version was like a chair and footrest or planchette with the rider sitting completely at right angles to the horse over which she would have had no possible control so her horse would have been led. Early versions of a form of side saddle appear as early as on Greek vases and carved on Celtic stones. This early version was introduced into England by Anne of Bohemia (1366-1394), the wife of Richard II. By the time of Catherine de Medici the stirrup had been added which took the place of the foot rest and the saddle had also become more practical, with the introduction of a pommel over which to hook the right leg, and a horn to secure the knee. However not all grand ladies took to them. Diane de Poitiers (1499-1566) the highly educated mistress of Henry II of France rode astride. Marie Antoinette figures in portraits sometimes side saddle and sometimes astride. There is a very fine painting of her in officer's uniform by Fournier-Sarlovèse. Catherine the Great (1755-1793) is also depicted in the uniform of a colonel of the famous Preobrazhensky regiment riding astride, by the Danish painter, Virgilius Eriksen.

But these grand ladies were more and more the exception. Riding side-saddle became *de rigeur* for ladies of the upper classes. Then in the 1830's a new addition was a second pommel called the 'leaping head' or 'leaping horn' which curved down over the left thigh making it much more secure. The left foot was slid into a single stirrup. The credit for this innovation is difficult to assign as there are at least two candidates Jules Pellier of France and Thomas Aldaker of England who had ridden side saddle as a dare and had a bad fall. A further improvement was the balance strap which ran from right rear to left front of the saddle. This meant that the girth did not have to be so uncomfortably tight.

In spite of all these redesigns the fact remains that it is far more dangerous riding side-saddle than astride as the horse is carrying a lopsided weight which must be extremely unbalancing especially when jumping. At the very least it tends to give the horse a sore back. Then if the horse should fall it is most likely to land on the heaviest side, making it just about impossible for the rider to jump clear, even if the pommels were not holding her onto the saddle. Apart from accidents, the twisted position is fatal to weak or injured backs of the rider, and can lead to curvature of the spine and the prolapse of the right kidney.

Queen Elizabeth I varied her riding arrangements from sitting behind the Earl of Leicester on public occasions, to side saddle for reviewing her troops and to riding astride when hunting. Queen Elizabeth II is a great rider and rides astride normally but when younger she attended Trooping of the Colour riding side-saddle until 1987 after which she rode in a carriage. In Spanish fiestas girls can still be seen riding pillion on special saddles in their frilly spotted dresses and behind their young men.

Another grand marriage of animal and technology was that of adding the stirrup to the saddle. Coming a lot later than the saddle, it was another of the most important innovations of all time. It is even considered to be one of the three great equine revolutionary inventions together with the chariot and the saddle.

Leather loops for the big toe were in use in India, even before saddles, possibly in 200BC and can still be seen in use in parts of South India and Malaya. Working with dates from so long ago can only be approximate and they are always subject to change as new facts emerge. Although leather stirrups may have been in use in Assyria, rigid stirrups arrived only in the

C4[th] from NE China during the Jin dynasty (AD265-420) and it was the Mongols under Attila the Hun (the 'Scourge of God') who used them in about 500AD thus transforming riding and enabling the nomads to wage war from horseback. The rider was not only more secure and comfortable in the saddle but he was then able to develop such skills as shooting from powerful bows even at full gallop. He could also stand up in his stirrups to use his sword or lance more effectively, with stronger thrusts. And as an extra, he could also use them for mounting. Before the stirrup, you needed to get given a leg up by another man or if dextrous, you could use your spear like a pole-vaulter and leap into the saddle.

By 600AD the stirrup had been introduced into Europe and came into general use during the next century or two. Curiously enough the Irish, as late as the C12[th], were still not using the stirrup generally. For Charles Martel's cavalry the use of the stirrup was considered so important that he rewarded those riders who brought to his army a horse, saddle and stirrups, with a title to church land. And thanks to this well equipped cavalry he was able to rout the invading forces of Abd el Rahman al Ghafiqi and the Ummayads in AD732 at the Battle of Tours or Poitiers – the battle was waged between the two towns. This was the battle that saved Europe from the relentless spread of Islam from the Iberian Peninsula at that time. Martel was really the introducer of cavalry to Europe.

The horse too gained a great innovation in the form of the *hipposandal* in 70AD, which consisted of a metal shoe tied to the hoof with leather straps which must have been quite tricky to attach. This was a primitive forerunner of the real horseshoe, The Romans used hipposandals, and so did the early Asians and in Japan straw slippers were used on horses right up through the C19[th]. However, by the C6[th] and C7[th]AD Europeans had begun to nail metal ones onto their horses' hooves. By 1,000AD cast bronze was used, and by the C13[th] and C14[th] iron was the chosen metal. But iron was valuable in those days and old horse shoes were used as a form of currency. And so the art of the farrier grew and grew as it became ever more necessary to shoe horses. Why had horseshoes become essential? Horses, in their original wild state move and graze slowly and continuously over dry hard ground, but when domesticated they are often moved to wetter heavier soils causing the hooves to become soft and begin to split.

The shoes are fitted after a form of manicuring to reduce the thickness of the hoof which would have been worn down in natural conditions.

Another accoutrement for man was the spur, part utility part dressing up. The earliest form was the 'prick spur', which was sewn onto the heel of rider's boots.. They lent themselves to the mediaeval love of display and status. The metal spurs that developed later consist of a heel band, which is strapped onto the boot. A shank or neck protrudes from it sometimes, ending in a rowel which is a form of wheel with spikes. Some spurs have a spike that fits into the heel of the boot. They were found in Celtic tombs from C5th BC and even mentioned by Xenophon at about the same time, in his manual on hunting.

In mediaeval times the knights' spurs were gilded and their squires' were silvered and the famous quote "To win his spurs" or "To gain his spurs" meant a man had gained his knighthood. In the event of disgrace the knight had his spurs hacked off.

There was a famous "Battle of the Golden Spurs" 1302 at which the French chivalry suffered a humiliating defeat at the hands of the Flemish. Infantry won over cavalry and thousands of golden spurs were collected off the battle field and hung in the Church of Our Lady in Kortrijk.

Even today at the coronation of a British monarch the 'spurs of chivalry' are part of the insignia. In mediaeval times, everyone wore them and the noise in church of the men clanking round could drown out the preacher.

A final bit of equipment for people dealing with horse is the whip. For riding and hunting the riding crop was used, consisting of a cane covered in leather ending in a leather loop with an L-shaped handle often of bone. For 'whipping in' hounds the crop has a long thong attached to the loop. The thong can be cracked and the sound is good for holding back hounds or to direct livestock. Dressage whips are flexible and quite a bit longer than crops – up to 110cm and a long lash for training. Whips can be used to inflict pain in extreme circumstances but are usually more like an extension of the human hand and arm to signal with a tap. When riding side-saddle they replace the right leg of the rider for signalling. In Australia the stock whip has been developed with a thong up to 3 metres in length. Similarly the coach whip had a long lash to reach the front horses of a coach and four. Great skill and practice was needed to handle the coach whip.

With all these inventions it became easier and more comfortable to ride the long distances required by war and peace. One of the great coincidences is how well man and horse fit together, physically, almost as if designed for each other. Charles Darwin certainly agreed, as while travelling in South America he remarked on a gaucho riding by. "A naked man on a naked horse is a fine spectacle; I had no idea how well the two animals suited each other." And certainly it is true that the horse is not too wide to ride astride comfortably; the saddle fits well into the hollowed back, behind the withers; the man's legs are neither too long nor too short to bend into a position to control the horse; the heels are in just the right position to urge the horse on by kicking and other signals. Also the reins running from the bit to the hand are in a straight line for reining in and other controls by hand. And compared to a camel or worse still an elephant, it is not too far to fall in case of an accident.

And so the Horse, ridden or as a draught animal was now of paramount consequence for travel overland and was to remain so until the takeover of the motorcar, the train, and later still the aeroplane.

Now horses were being bred to be faster and stronger. Harness was invented to make it easier to manage them but another really important aspect of owning and riding a horse is the training. Soon after 2,300BC the Sumerians had written manuals on training horses and even pedigrees. But one method of training stood head and shoulder above all the rest. In 1,350BC, in Central Anatolia the Hittites produced a man (actually a Mitanni, from the Tigris/Euphrates area) called Kikkuli. He gave his profession as 'assussanni', which means a procurer or gainer of horses. He inscribed his methods in cuneiform on clay tablets, which luckily have been preserved and so we know exactly what his technique was for training a horse. He had to combine getting it fit and at the same time avoid it injuring itself. He reckoned a horse needed the equivalent of 100 miles of gallops before it was fit enough to really race. The seven-month training regime is particularly amazing because of Kikkuli's instinctive knowledge of the physiology of the horse. This was over three thousand years before George Stubbs made his famous studies of the anatomy of the horse. To maintain the Hittite stable, huge staffs of trainers and grooms were necessary; food bills had to be paid; and all the equipment had to be fashioned and maintained. But Kikkuli had produced so sound a routine, right down to details about food and water, that in 1991 the 3,300 year-old techniques were actually resurrected and used in Australia at the University of New England by a Dr Hyland with excellent results.

One thousand years later in 322BC no lesser person than Xenophon (430-354BC) wrote 'The Art of Horsemanship'. Xenophon was an all rounder, a real polymath, being amongst other things, a literary man and a soldier all combined with having an intense interest in horse training. He had first gained experience serving in the Peloponnesian Wars under Cyrus, son of Darius II where he had picked up tips in the various uses of horses from cavalrymen from other countries. For example, from the Persians he learned the trick of being lifted onto a horse by being given a 'leg up' by another man. By the Armenians he was shown how to protect his horses' hooves from ice and rocks by tying pieces of cloth round them. From all this knowledge that he gained, he gave details for the care and training of horses both for military and civilian use. Curiously, one of qualities he recommended in a horse was a fleshy back though no doubt this was prompted by a thought for the comfort of the rider in the days before saddles. One excellent piece of advice was "The one great precept and practice in using a horse is this, never deal with him when you are in a fit of passion." He recommended a gentle but firm approach to training, grooming and caring for a horse. He emphasized the importance of communication between horse and rider to create a partnership. His excellent advice was followed to this day by much of the equestrian world but was unfortunately ignored by some of the trainers who felt that 'breaking in a horse' meant literally just that. One rather good piece of advice was that the rider should sit on a horse not as if in a chair but as if standing with the legs apart. He also detailed the principles of classical dressage and rather charmingly mentioned the thrill of seeing his horse curvet.

Owners of horses have to realise that they like routine and so taking a horse to war is extremely upsetting. Even the best-trained horse is unpredictable. Every rider knows how you can ride nine times past some object and then the tenth time the horse can decide this object has suddenly become a thing of terror and shy all over the place. Another strange thing about a horse is that they really feel happiest and least vulnerable in their stable. When you take a dog for a walk, the farther you go the better, but when you ride out, the horse really gets keen when he knows he's heading for home. This vulnerability must have come from their selective breeding where the emphasis was on tractability and compliance and of course, in the wild, flight is their first instinct when faced with danger.

In 60AD a book on agriculture called 'De Re Rustica' was written by a man called Lucius Moderatus Columella (AD4 – 70) in twelve volumes which have survived to this day and covered every aspect of farming from viticulture to bee keeping, from gardening to breeding stock. Volume 6 deals with Cattle, Horses and Mules. The mares producing mules were particularly highly valued. It may be just a coincidence but Columella was born in Spain and by the C18th and C19th Spain was breeding by far the best mules. He describes how the jackass that had been selected for fathering mules had to be kept with horses to adapt to them and respond to the mare when she came into oestrus. They even built a special mating ramp for use when the donkey sire was much smaller than the horse dam. For horses he recommended daily grooming and stated that most ailments were caused by cold, fatigue and drinking when the horse was too hot. He also recognised swimming as a cure for lameness. An obvious observation was that better food meant bigger and better horses. He also advised a careful selection of mares and jacks to select the best traits for the breeding of mules. All of these views are held again today.

From the C11th onwards in Mamluk held territory (Syria, Palestine and Egypt) there were written many treatises called Furusiyya, an Arabic term for horsemanship, and more particularly 'knightly martial exercises'. The three basic practises were horsemanship, archery and charging with a lance, but it did include care of the horse and veterinary practices. The Mamluks were being threatened by the Crusades and needed to upgrade their cavalry.

Claudio Corte, an Italian, also wrote a book in 1584 recommending kindness and patience and rewards but also punishment for bad behaviour such as a horse that kicks out should be whipped. And he also suggested sham fights to get horses accustomed to weapons, and swimming for getting them fit.

A more modern trainer of horses was François Robichon de la Guérinière (1688-1751) who also advocated a firm but gentle hand. His book "Ecole de Cavalerie" is one of the most important manuals of horse training and was used and still is, by the Spanish Riding School. He was most insistent on the rider having a good seat and therefore a soft light hand. Punishments were to be light and few. His aim was a fit supple horse.

A fellow writer of a training manual was William Cavendish, the first Duke of Newcastle and the C17th owner of Bolsover Castle who wrote "A New Method and Extraordinary Invention to Dress Horses and Work them according to Nature" (1667) Cavendish was a polymath who added architecture to his other talents of poetry, politics and horsemanship. He built the castle's Riding House which housed the horses and provided an indoor riding school. His book was written rather on the lines of the Viennese Spanish Riding School with the horses being taught elaborate caprioles and dressage. A famous quotation of his is "There is nothing of more use than a horse of manège (training of horses); nor is anything of more state, manliness or pleasure, then riding".

In 1833 Darwin met General Rosas in Brazil. He had been elected as general by a strange and dangerous trial: a troop of unbroken horses was driven into a corral through a gateway under a cross bar: and the man who could drop onto the back of one and ride without saddle or bridle until it was subdued became the general. So the General was simply the best and strongest gaucho in a country where skilled riding was essential.

Darwin goes on to relate the ruthless training methods of gauchos – a far cry from the gentler methods of Kikkuli and Xenophon. A *domidor* (trainer of horses) catches a wild horse colt in a corral with a *lazo* or rope round its front legs. The Horse falls to the ground, and the gaucho ties one back leg to the front ones. Then he sits on its neck and bridles it, but without a bit, then he unties the back leg allowing the horse to rise with difficulty. A second man helps put on the saddle. The wretched horse throws himself over in fear again and again. "The poor animal can hardly breathe from fear, and is white with foam and sweat" (Darwin). As the gaucho mounts he pulls the slipknot from the legs and the horse is free. He gallops until the horse is exhausted and can be brought back to the corral. In two or three weeks he becomes quite tame or subdued. Later a bit is introduced as he must learn to associate the will of the rider with the rein. Darwin remarks that animals are so abundant that "Humanity and self-interest are not closely united". He was actually shouted at by the owner when he tried to rest an exhausted horse he was riding.

In the mid C19th an Englishman called Richard Darvill literally rode to the rescue of the art of training. He had run away from home and first became a stable boy, then a veterinary surgeon and finally joined the Hussars. He was shocked by the appalling ignorance prevalent amongst horse owners and the grooms. So he wrote his own carefully researched and very precise notes, which he dedicated to the Jockey Club. He found there were some terrible instances of abuse, frequently out of sheer ignorance. For example one owner neither fed nor watered his horse for twenty-four hours before a race. Darvill went into everything from stabling, feeding, illnesses, ventilation, clothing, and shoeing. For him, the horse's well-being was paramount; he advocated little bribes of corn to sooth or to distract a nervous animal while it was being shod or mounted for the first time.

To accompany their riders, horses began to be exported all over the world. And so at long last horses made their return to America, their country of origin thousands of years after they had left it. This happened in 1494 when Christopher Columbus landed with his army accompanied by twenty-four stallions and ten mares...a very long time since they had crossed the Beringia Land Bridge – 136,000 years before. This was only the start and more and more horses were shipped over by the other Conquistadors. Horses were introduced into Mexico in the 1500's. Following the Pueblo Revolt of 1680 when the Indians drove the Spaniards out of New Mexico, horses spread very rapidly throughout North America. So by the 1700's the horse was right back in Colorado in time to play Cowboys and Indians. The Indians, who had no record of ever having known horses couldn't believe their eyes at first and called them 'big dogs'. The sight of them became an instrument of terror in its own right. Then the Indians began to take them on. Some of the Amerindians became interested in selective breeding. An example is the Nez Perce tribe who bred Appaloosas, wonderful spotted horses who transformed the tribe from sedentary fishermen into hunters and traders with the new power gained from their new mobility. Apart from their markings the Appaloosas were bred for temperament and intelligence but unfortunately the Nez Perce wars of 1877 led to dispersal and slaughter of the breed. But by the 1930's they were being bred up again under the direction of the Appaloosa Horse Club of America. Back in Europe, in Vienna, the Appaloosas were being used by the Spanish Riding School.

Shipping horses is always hazardous as apart from risk of injury in rough weather, they are unable to cope with sea sickness, so nearly half the horses died at sea. As the numbers of horses built up, the inevitable happened, some got loose and ran wild. Feral horses in America are known as Mustangs and by 1900 they had bred up to an estimated two million. Their numbers were drastically reduced by methods so inhumane that in 1971 a protection law was passed.

With the advent of colonisation horses were taken to other far flung places such as South Africa, shipped there by van Riebeeck in 1653. They were a necessary part of the Dutch trying to establish a refreshment station for ships rounding the Cape en route to the Far East for the Dutch East Africa Company. The three month voyage from Europe was even longer than to America and after landing the next hazard was the deadly horse sickness, carried by tsetse flies. The shorter route for supplying horses was from Asia. South Africa's horse history had begun with Barbs and Arabs being imported and then a breed was established with a mixture of blood from thoroughbreds. These horses were selected for hardiness and endurance and were so taken over by the Boer farmers that they became known as Boereperd (Farm horse). An offshoot of the Boerperd is the Basuto pony which was bred from stolen horses and adapting to harsh conditions in feeding and terrain grew smaller and stockier and became used not only by farmers but for trekking across the mountains of Basutoland (Lesotho). The scourge of horses in South Africa was and still is horse sickness, spread by midges in the hot rainy season. Zebras are immune to the disease, but horses, mules or donkeys that catch it have the terrifyingly high mortality rate of 70% to 90%.

The horse had even further to go to Australia and the first nine horses were introduced into Australia in 1788 with the arrival of the first fleet into Botany Bay. Hardy and sturdy were the desirable traits which came from a mixture of Spanish horses and thoroughbreds. Later arrivals were Welsh ponies and Arabian horses and this mixture of breeds were first named 'Walers' and then termed Australian Stock horses. The very traits of strength, endurance and good temperament for use on the Australian sheep stations also made them ideal for going into battle, so back to the popular old game they were taken and used in the Boer War in South Africa, WWI and even by the Indian cavalry.

The next thing horses and especially horse-drawn vehicles needed were proper roads. During the mediaeval period the Roman roads had deteriorated into tracks; then by the late C17[th] they were making a come back and being used by the stage coaches for the public and also the carriages of the wealthy landed gentry. In Paris in 1662 the mathematician and eccentric inventor Blaise Pascal (1623-1662) devised a bus service, with buses pulled by horses, but unfortunately it never really caught on. It was probably before its time.

Much later in the early C19[th], a coach builder named George Shillibeer (1797-1866) was invited to Paris to design and build a horse-drawn coach "of novel design" to carry up to two dozen people. Back in England he produced something similar for the Quaker Newington Academy for Girls. Then in 1829 in London the first horse-drawn "Omnibus" started operating a public service between Paddington and Bank, four times a day. This omnibus was very broad so very safe and was drawn by three horses abreast. These services grew and grew but first there were no designated stops you just hailed the bus anywhere along its route. Every town in Europe depended on horses for every sort of communication, provisioning, transport. By 1789 there were estimated to be 21,000 horses in Paris alone, and this produced trouble of another sort – what do with the tons of manure produced every day.

Before the Industrial Revolution, blacksmiths/farriers were essential in every village in England. They not only shod all the ponies and horses and made and repaired household and farm implements but seemed to be the centre of gossip. They even sometimes extended their strength and expertise to a rough form of dentistry; it must have been very rough. By the C14[th] farriers were also acting as forerunners of vets. The Worshipful Company of Farriers was formed in 1356 but unfortunately, its early records were lost in the Great Fire 1666.

After Chariots, as we have mentioned, came carriages and coaches. Like everything else that enters our lives there was an enormous range of transport from the common cart to the carriages of the gentry. For example at the top end of the range was the Landau which was the preferred fashionable carriage but as it was open, it could only be used in fine weather. The other upper class carriage was the fast and dangerous Phaeton which was the Ferrari of its day. For public transport there was the stage coach, which went all over England. A transport service was begun in 1785 from London to Edinburgh, a journey of 400 miles, which took ten days in summer and twelve in winter. The first horse drawn tram appeared on the Isle of Wight in 1876. In the USA the first Concord Stage Coach was built in 1827 and the use of leather strap braces made it better sprung. Over 700 were built and supplied to South America, Australia and Africa. Passengers were crammed in, in acute discomfort and also had to endure possible attacks by highwaymen and Amerindians.

Horses, mules and at the bottom of the social scale donkeys ruled the transport system from one end of the world to the other for over six thousand years. But then the first man waving a red flag came over the hill signalling the advent of the first motor car and the world began to change again. Horses are no longer essential to the transport in the Western world, but in undeveloped countries, donkeys, oxen, camels and llamas are vital modes of transport.

ET CETERA

1. Far from the glamorous Boudicaa usually depicted heroically driving a chariot drawn by great thoroughbred looking horses, with wheels armed with jutting knives, the reality would probably have been a tough and dirty woman in a small rough cart drawn by little shaggy ponies.
2. Just a few famous mounts from literature are Black Bess, the amazing Mare created for Dick Turpin by Harrison Ainsworth for the great ride from London – York. Then there was Rosinante, Don Quixote's ancient hack, all skin and bone and Sancho Panza's little pony Dapple.
3. Every church, every inn, every big house had its mounting block or horse block to enable riders to mount without effort. In Scotland they were called a louping-on stone. They were of particular use in the days of heavy armour.
4. Then there were the Horse Latitudes – the region of calm around 30degrees north and south of the equator where sailing ships carrying horses on their way to America or the West Indies were sometimes, when becalmed, forced to jettison their animal cargo, owing to a shortage of water.

CHAPTER 5

SHIPS OF THE DESERT & THE MOUNTAINS

Ships of the Desert and Mountain...Camel at Pushkar Camel Fair

"This is the camel of Allah, a token unto you, so suffer
her to feed in Allah's earth, and touch her not with harm
lest a near torment seize you."

Koran 11:64

Now for the story of the camel, who was to become essential to man in so many parts of the world, and this too is one of the most fascinating of all histories. Like horses, all *camelids* originally arose in North America.

Their ancestors wandered the plains at least 40 million years ago. And they too crossed the Bering Land Bridge into Asia about 3 million years ago. However there was another branch of the camel family that gradually migrated south about 1.4 million years ago and established itself in South America. These ones slowly evolved into the ancestors of the guanaco and the vicuna, from which were bred the domesticated versions the alpaca and the llama. So by the end of the last ice age (10,000 - 12,000 years ago) *camelids* were extinct in North America. Man certainly had a hand in their extermination – killing them as usual for food. But why they totally vanished is another one of the great mysteries of pre-history that no one has yet solved. The antecedents of the camels and llamas were no larger than hares and possessed four toes. Fossil footprints of this 'Procamellus' have been found in California and even actual fossils of tiny creatures the size of rabbits. North America has this amazing long fossil camelid record ranging from the three foot tall *Genticamelus* to one 17 feet tall (*Aepycamelus*). One of the most common was the rather charmingly named 'Yesterday's Camel' (Camelops hesternus) which was a long-legged version of the dromedary.

From the ancestors that had crossed the Bering Straits two camels developed – the Bactrian (*Camelus bactrianus*) with two humps and the Dromedary (*Camelus dromedarius*) with one. As a seventeenth century traveller succinctly put it: "Providence has made two types of camel, one for hot countries and one for cold." Here were our next modes of transport. They are probably not two different species of camel but geographical races of a single species. This theory is borne out by the impossibility of telling the difference between them from their bones. The curious thing is that the two varieties can be interbred and the hump trouble gets ironed out into a more or less continuous one with a cone at each end. The other useful consequence is that their performance is improved by hybrid vigour, that same mysterious power that emerges when horse and donkeys produce mules.

Fossils have been found as far apart as Algeria, Kashmir and India. And a more recent discovery reported by the National Geographic News, of October 11, 2006 was of the skeletal remains of a giant extinct camel near Palmyra in Syria. It was estimated to have been twelve feet tall, double the size of the present day camel. The Swiss-Syrian team of scientists estimated that this species of camel lived about a hundred thousand years ago making the find 90,000 years earlier than originally estimated. Human remains and tools found nearby suggest that these enormous dromedaries were hunted by Man. In that period the region would have been savannah not desert.

But the big question is when and where were camels first domesticated? According to The Bible it was about 3,000 – 2,000 BC, the age of the Patriarchs. In Genesis 24 Abraham sent his head servant, Eliezer, to Mesopotamia with ten camels to find a wife for his son Isaac, and Eliezer found Rebekah who watered and fed his camels. Also in Genesis 24:10-11 and in other verses the importance of camels in Abraham's life are shown. Another patriarch was Job, more or less contemporaneous to Abraham "His substance also seven thousand sheep, and three thousand camels, and five hundred yoke of oxen, and five hundred she asses…" They are also mentioned vividly in Judges 7:12 "And the Midianites and the Amalekites and all the children of the east lay along in the valley like grasshoppers for multitude; and their camels were without number, as the sand by the sea for multitude." Those tribes were probably in the area south and east of present day Palestine.

Camels are also mentioned in connection with the Queen of Sheba in II Chronicles 9:1. In the C10th the Queen of Sheba learned of Solomon from Tamrin the leader of her camel caravans. Tamrin owned his own great retinue of camels and donkeys and had visited Jerusalem many times. The Queen was so impressed with his descriptions of Solomon's wisdom and perhaps also of the Temple Solomon was building that she set off across the Arabian Desert, from Ethiopia, a distance of nearly 1,500 miles to visit him. Her caravan of 797 camels was laden with precious gifts, spices, gold and precious stones. The journey must have taken months as camels can manage only about 20 miles a day and any baby camels born along the way would have been loaded onto their mother's backs. Sheba herself rode in a magnificent golden palanquin. The visit was more than consummated and gave rise to the birth of Prince Menelik to the Queen of Sheba whose bloodline continued down to the present time until the overthrow of Haile Selassie the last Emperor of Ethiopia in 1974.

In early 2014 strong opposition flared up between evangelical scholars and the archaeologists who are basing their dates on recent excavations of camel bones that were found, near ancient copper mines in the Aravah

Valley, south of the Dead Sea. This date, according to carbon dating, points to about the end of 1,000BC, a good thousand years later than the biblical date. However these excavations do not rule out the possibility that camels could have been domesticated earlier, although the Bible bases its dates on belief rather than on facts. So a firm date is difficult to come by.

Of the Asian camels the dromedary (from 'dromus', Latin for road) was probably the first to be domesticated on the Arabian Peninsula towards the end of the 2ⁿᵈ millennium BC. It is now extinct in the wild. The other camel, the Bactrian, was domesticated in Asia and got its name from Bactria, the Greek name for the Oxus valley in North Afghanistan, the original homeland of Aryan tribes who had moved into Iran the home of the prophet Zoroaster (dates for his life vary wildly from C18ᵗʰ to C6ᵗʰBC). There are still some in the wild; they are very much on the endangered list.

The earliest known representation of a camel are two carvings near Jabal Tubayq in eastern Jordan, dating from the Stone Age; but the first concrete proof of domestication is a picture of a camel with a rider found in the ruins of Tal Halaf in Iraq, dating back to at least 1,000BC. The first known actual reference to the one-humped Arabian camel is a pottery figure from Egypt, which also dates back to about 3,000BC. Two early stone-age carvings of the single humped variety have been found near the Eastern border of Jordan at a place near Abu Tubayq. By the 1ˢᵗ millennium BC camels were depicted on Achaemenid Persian and Assyrian bas-reliefs. The Apadana Palace in Persepolis, founded by Darius the Great in 515BC, shows a huge bas-relief on the terrace walls of a delegation bringing two-humped camels from Bactria. They also appear in some of the manuscripts in the famous sealed Buddhist library of Dunhuang, on the Silk Route (dating from C11ᵗʰ and found in 1900).

Camels appear vividly in Chinese art. The Chinese of the Han (206BC – 220AD) and T'ang periods (618AD – 906AD) had a tradition of ceramic tomb sculpture, also known as 'mingqi', which were supposed to provide for the deceased in the afterlife, so camels and horses too were depicted sometimes pulling carts or ploughs. These ceramic sculptures recalled the wealth of luxury goods that both reached and left China along the Silk Route – an outward signal of prosperity.

The Nabateans depended entirely upon camel caravans for moving their goods. They were famous for building Petra in Jordan, famously described in a prize winning poem by John William Burgon as that "rose-red city, half as old as time" Pictures of camels – almost classed as graffiti, are found painted all over their *wadi* walls.

Both varieties of camel have evolved the most amazing physical features to cope with the extreme conditions in which they live. Starting with their distinctive humps – the popular misconception is that camel's humps are used for storing water. This is not so, water is stored throughout their bodies. The humps store fat, and provide a built-in food supply in times of need. The dromedary has one hump whose hump just shrinks, when short of water and food unlike the Bactrian with two which tend to flop over when fat is lost, as is shown so clearly in the pottery figures of the Tang Dynasty (618-907). How the camel got its hump has been the object of guess work and legend and the Rudyard Kipling story in which he got his hump because he was too lazy to work along with the horse, the dog and the ox – three of our domesticated animals. However recently a team in Canada discovered the fossil of an 11 foot ancestor which inhabited the Arctic Circle millions of years ago and reckoned the hump of fat was storage to get them through the winter.

The one humped camel or dromedary was domesticated in the Arabian Peninsula about 3,000BC or even earlier. It weighs about the same as the Bactrian 300-600kg but is taller at 215cm to the top of the hump, and covers about 40km a day, unless it is a specially bred racing camel. Another feature is their double row of long eyelashes, which protect their eyes from glare and blown sand and a further useful quality is that they can close their nostrils at will. They have very small ears with thick hair to keep out the ever intrusive sand but very good hearing. When short of water their urine can be as thick as syrup and holds double the amount of salt as sea water.

The cold climate camel is of course the Bactrian; being adapted to the cold mountains they have longer darker shaggy hair and thick under wool, which insulates them against the extremes of winter cold (–30 deg. C.) and summer heat (+40 deg. C.) Standing over 2 metres high and 3 metres long they have two humps and weigh over 700kg. They also possess shorter legs and a more massive body. Another difference from the Dromedary is that they can cope with rough mountain trails and have also developed various devices to deal with snow and sand. Amongst their special characteristics are their feet that splay out just like snowshoes, which prevents them sinking in. With their tolerance of cold, they are splendidly adapted for freezing deserts and mountainous regions. In the Pamirs and Tienchan they can manage altitudes well up to 4,000 metres, above which yaks are used. They can cover 50 km a day, carrying loads of 130-200kg, twice as much as an ox over terrible mountain terrain. A caravan of 1,000 camels could carry 400 odd tons, the equivalent of a sailing ship. The Bactrian is indigenous to Mongolia, the Iranian plateau, China and the Gobi Desert. It was Aristotle who first used the name 'Bactrian' as opposed to 'Arabian'. In about 2,500BC these camels began to be domesticated in the area after which they are named. Their Northern limit is about 52 º Latitude into Siberia, Mongolia and Northern China. Even now there are still a few wild herds there, but tragically this magnificent animal is on the brink of extinction in the wild, owing of course, to hunting and is amongst the world's most endangered mammals. Originally Bactrians were used in the Syrian Desert about 3,000 years ago but their coats were too thick and hot for local conditions so they were replaced by dromedaries. Camels were and still are used in the central and Eastern parts of Asia such as the Gobi Desert; so their nickname of "Ship of the Desert" is well earned.

Camels are exceptionally economical feeders and because they keep moving they don't overgraze like sheep and more especially goats. They have several amazing adaptations to fit them for the desert life, as we have seen with the Bactrian, they conserve water throughout their bodies and there are pockets in the first two of its four stomachs where water does get stored; desperate Arabs have been known to slit open a camel's stomach to get this water in extreme circumstances. They can go five to seven days without water and then can drink over twenty gallons in just a few minutes. Another water conserving adaptation is that they also store water uniquely in their bloodstream; they also produce very little urine and extremely dry faeces (which can actually be used as fuel right away). If a camel refuses to eat it is a sure sign they need water. Their hump actually wastes away when the hydrogen stored in it is released to combine with oxygen to make water and when that is gone the camel dies. Apparently in dire desperation tea can be made out of strained camel's urine. The desperation would have to be pretty dire.

In winter when they feed on juicy green plants they can actually manage without drinking for up to fifty days, but in summer they can't manage more than five and really prefer to drink every couple of days. The Bedouin sing to soothe them and encourage the water down. Because the camel's lips have thick hairs protecting them they can eat tough thorny shrubs that even sheep and goats cannot manage. Other modifications are their eyes which have long eyelashes to keep out the sand and are specially adapted to bright light and also possess semi-translucent eyelids through which they can see the way when closed during a sandstorm and they can shut their nostrils at will. Their fur acts as insulation and keeps them cooler as a covering against the sun which they always have the sense to face to minimize exposure.

A camel's blood temperature can be raised 6° without ill effect; also because it doesn't sweat it conserves moisture. 12% loss of body weight of water can kill a man, but a camel can survive a 40% loss. Animals in the cold tend to huddle together to conserve warmth but camels huddle together for the opposite reason – another body is cooler than the outside temperature. When they are desperate for food, camels have been known to eat meat and even bones and fish. Camels are bred for pack, riding, farm work, milking and as we see later for racing. Unlike the horse or the dog that moves alternate legs, the camel 'paces', which means that is moves both legs on the same side in the same way as giraffes and elephants. For these animals the reason for this gait is that they have long legs quite close together and if they moved like a horse or dog they would seriously overreach, which means the back leg would come into contact with the front leg and cause injuries. Unlike the other ungulates that have hooves, the camel has big soft pads split in two with toenails.

Dromedaries have gradually moved west across the dry belt of Eurasia and some went south into India, S. Russia, Nubia, S Arabia, Middle East and North Africa. At present in North Africa there are about 20 breeds and the main home is the Northern Sudan. Fossils have been found in Morocco, Algeria, North Kenya, Ethiopia and Somaliland. But by historic times the wild version had become extinct. In Asia, they are in Turkey, Syria, Lebanon, Israel, Jordan, Iran, Iraq, Afghanistan and Pakistan. By C2[nd] they were common in Northern China as the Chinese also began to realise their value.

Camels arrived about 2,000 years ago in Somaliland but actually from S Arabia and not from N. Africa so perhaps they were introduced by the seaborne incense trade and there are rock carvings on Socotra dating from C10[th] to prove it. Amazingly, for their sea passage they were fed on dried shark & sardines.

To the medieval Muslims, the most important of all the animals was undoubtedly the camel because Islam controlled the bulk of the inland routes and the camel was the absolute mainstay of its trade. They were called "the Gift of God" as well as "Ships of the Desert" and were absolutely vital for the great trade between East and West across Eurasia. The "Ship of the Desert" metaphor came from the Koran and Bedouin poetry. All the Trade Routes through central Asia and Arabia, such as the Incense Route, the Silk Route and the Tea and Horse Route owed their very existence to the camel caravans, whose tracks were built by them and for their use and not for carts. Even today modern trucks and 4x4s cannot always follow where a camel goes.

They were also used to convey pilgrims from all over Islam for the Hajj or Pilgrimage. This pilgrimage is the fifth pillar of Islam, and is a religious duty that must be carried out at least once in a lifetime by every able-bodied Muslim who can afford to do so. Some caravans consisted of up to 2,000 camels long. To encourage trade and accommodate pilgrims the local kings built caravanserais or camel hotels at intervals of about twenty miles all along the trade routes. Apart from being a very long journey, it can be extremely hazardous journey; for example in the C17[th] 1,500 people and 700 camels died of thirst on the way to Mecca.

One third of the earth's surface is arid, (the definition of which is that evaporation exceeds rainfall) and one third of all this dry land is the Sahara desert, which is the hottest and driest of all deserts and can reach temperatures of C70° and is especially bad in the rocky areas, which retain the heat. It seems amazing to think it was once savannah, as is shown by the animals, especially cattle, in its rock art. Only about 10% of the land is fit for cultivation for humans – dates and peanuts and a little wheat. So it must have been a godsend when camels arrived there about 2,000 years ago from Western Asia. The Libyans, ancestors of the Tuaregs used them to make quick raids on the Romans and retreat back into the desert.

By the seventh and eighth centuries the Arab conquests had introduced many more camels. The irregularity of rainfall makes the nomad life absolutely essential.

The Sahara nomads are very proud of their independence. Their powers of tracking are almost magical as they can tell everything about the camel's condition and history from just a few footprints. The bush telegraph works so well that word gets around over miles of desert on the whereabouts of pasture and water. Camels can find their way, no one knows how, across great tracks of desert and not surprisingly, they can also smell out food and water from a great distance. They also home in on familiar territory. Camels are so adapted to withstand extremely harsh physical conditions that they, in fact, do not prosper in urban conditions or rich pastures and in damp climates they can be susceptible to disease.

Other curious features of camels; they are not frightened by gunfire, which makes them useful in warfare, but are surprisingly nervous of strange people or sudden movements and in Australia they discovered that like elephants they are terrified of pigs. But like everything else the camel has its drawbacks – they are rather stupid, which makes them difficult to train; they are prone to bad temper, especially the males who can be really aggressive, especially in the breeding season, and their smell is most unpleasant particularly to horses and donkeys, who, when first encountering it, can be panicked by it; the Romans kept them well outside their posts on account of that trait. Another disadvantage is their slow breeding rate, which only starts at the age of five. A calf is born only every 3 or 4 years. They can't be penned, so not popular where the ass or ox can go, but a boon to the nomads, who have an amazingly close relationship with their animals. When the heat becomes very intensive they like to roll on their backs in soft sand or earth, perhaps to soothe and cool themselves. For Islam they fulfilled every role of transport and were even used for tilling the land, turning water wheels and pulling carts.

Camel milk is essential to the nomads and the saying "There was plenty of milk," means it was a good year. Camels provide frothy light milk, which is made into delicious cheese. The Bedouin drink it more than water, and it has more fat, protein and Vitamin C than cow's milk. In Kenya the milk is mixed with blood drawn from the jugular by shooting in an arrow at close range. They are also a source of food, not only meat (pretty tough) but the fat from the hump can be made into a sort of butter. Camels are slaughtered mainly for feasts, such as weddings. On these occasions it is customary for the bridegroom to present his new father-in-law with fifty female camels. For murder, the perpetrator gives the victim's family one hundred camels, unless the victim is only a woman, whereupon the reparation is halved!

Called *tiktar,* camel meat can be dried in the open air rather like South African *biltong*. For the nomads, the camel's main use is for carrying water – 100 litres, and moving camp. The usual pace is only thirty kilometres a day combined with feeding and rest but they can keep this up for weeks. The combination of their sense of orientation, their observation of the tiniest detail in the terrain and their fatalistic belief in Allah will bring them to their destination every time. But it is endless work just staying alive. Nevertheless the nomad life should be encouraged as only nomads with their camels can make full use of the Sahara as a resource. Like their camels they find it impossible to adapt to town life and soon drift into the slums as happened during the severe drought of the seventies. It seems all the government needs do is to organize and mend wells and water holes and leave the rest to the nomads.

Owning plenty of Camels was the ultimate prestige. They are the subject of poetry, proverbs and are compared to jewelry, beautiful women, and the finest weapons. The Arabs have over seven hundred words to describe a camel, (like the Inuit's extra vocabulary to categorize snow and the Mongols with their horses); for the coat colour alone there are over forty. As with Zulu cattle skins, white is the favourite.

To the nomadic Bedouin of ancient times camels and goats represented their means of transporting family and wealth. They were also central to their life on the spiritual and intellectual level, especially the Tuaregs and the Somalis. The latter have three or four million camels, which they use mainly for milk and as pack animals and never ride. The Horn of Africa holds the largest population in Africa. Millions of people in Asia and North Africa still depend on them for almost their every need. They pull ploughs, turn water wheels and are often the only form of transport. Other by - products are their soft warm wool, which can be woven into blankets and tents and clothes. Textiles woven from camel wool have been found in a royal tomb in Siberia. The leather is extremely tough and makes excellent saddles, bags and shoes. Nothing is wasted. Their bones have been carved for centuries and are now taking the place of ivory in the intricate carvings by the Chinese.

Riding camels is still of paramount importance. Their bridle arrangements are unlike anything to do with the horse and consist of a simple nose peg usually made of wood with a single rein, which you pull to one side then flick it over the camel's head to pull it the other way. It is not a very accurate guiding arrangement but luckily that is not exactly necessary in the desert.

However everything has its drawbacks and the obvious one with the camel is its hump. The horse's back is perfectly dipped for taking a saddle, but the hump on a camel is just where it would be convenient to sit. The hump cannot bear heavy weights without becoming misshapen. Anyway it changes shape all the time according to its food and water intake, and even the seasons. So the great question was – where to put the saddle? There were three solutions. Behind the hump meant that the saddle pad fitted behind but reached forward round the hump and was attached to a wooden bow in front of the hump tied on with a girth under the belly. This was the solution in South Arabia and in the Oman. In the Hadhramaut they still use these saddles. However trade was increasing and camels were becoming more and more popular and by the reign of Tiglath Pileser III (745 – 727 BC) the Assyrian king, they were being mentioned in really large quantities; so the camel saddle had to evolve too.

The next answer was the North Arabian Saddle, which appeared between 500 and 1,000BC. The saddle was placed over the top of the hump on a sort of doughnut shape underlying it. Sticks either crossed or straight, in turn connected two large arches or saddle bows and were fixed, one to the front of the pad and the other to the back. The whole rather convoluted construction led to a rigid square with the weight of rider or load evenly distributed. Compared to it a horse's saddle is simplicity itself. But its use led to the camel breeders now becoming very important economically and even politically. The Reguibat tribes living in the West developed another saddle – the *Rahla* or Mauretanian but it was really more for load bearing than riding.

In the Central Sahara they developed the third kind of riding saddle, the *terick* used by the Tuaregs. This saddle was strapped on in front of the hump, with the driver resting his legs on the camel's neck. This meant they could use their toes for directing it. In a wide-open desert you don't have to be too accurate with steering. Other advantages were a lighter weight, a single girth, and far less need for adjustments for the hump's varying size. As it happens the front legs of a camel are far stronger than the back ones so they took the weight better. However conservative attitudes meant that in the Sahara North Arabian saddles were still used for women and in North Arabia the Sahara the Chaanba tribe in the Northern Sahara have a special saddle called the *bassour,* which is draped with curtaining to hide the women from men's eyes and can contain a woman and several children. In the Western Sahara women don't hide their faces and have an open saddle, the *amshaqeb.* The exception to all these camel saddle fashions are the Somalis who, using their beasts mainly for milk and meat, had a very poor and inefficient seating plan, a sort of cobbling together of mats and crossed sticks. The Socotran saddle consists of a succession of thick mats over the hump and down the back bound with cords into a ridge, then long baskets on either side. The seat for a rider ends up a risky thirteen feet above ground.

Some people do not ride their camels but use them only as pack animals such as the Chaanba, a Berber people. The terké saddle in Iran and Pakistan is also only for carrying packs.

So the single hump on the dromedary poses all kinds of problems but at least with Bactrians the saddle being placed between the two humps easily solves the dilemma. The one-humped camel began to replace the Bactrian in Iran at the same time it was replacing the wheel elsewhere.

As far as tribal dromedary owners and breeders were concerned there were just two types of camel – common and thoroughbred. Thoroughbred camels came from a long line of selective breeding and were especially bred for warfare. The best thoroughbreds came from Qatar, the Oman and Mahra in Southern Arabia. The camel breeders were drawn from the noblest tribes, vastly superior to goat and cattle breeders. The female comes into season late autumn or early winter and the gestation period is a year. They bear offspring two or three years apart, so breeding takes time. Speed, size and strength were the main factors and looks only followed as a result. The riding camel is bred to keep up a speed of up to 10mph for as long as 18hours.

Hybridisation was practiced from early times in the 2nd half of the C2nd under the Parthians in the Tigris and Euphrates valleys. The preference was for male Bactrians being crossed with female dromedaries. The results were not infertile like mules but the breeders found it best not to breed from the hybrids. With typical hybrid vigour the offspring were larger and stronger than their parents and the hump could be single with an indentation towards the front or it could have a sort of semi-flattened hump, or even a long raised part with two small humps. These were much more fitted for pack animals, and could carry about double the weight a dromedary could carry and more than double a Bactrian could, indeed 400 – 500kg was normal. Strangely the offspring of two hybrids could be bad-tempered and of poor quality. So the chosen sires were always Bactrian and the male hybrids were usually neutered. Eventually selective breeding by man evolved a one-humped, longhaired, cold-resistant animal that was not a hybrid and could breed. By Islamic times Bactrians were used as breeding stock only, except in Central Asia where the climate is so harsh only Bactrians can survive and the Kirghiz are linked to a camel culture not unlike the camels of the Sahara nomads, where camel milk became a very important part of their diet.

Dromedaries were only introduced into India after 1,000AD. Then they predominated over the Bactrian camels which were mainly used in the high mountains. There are three major Indian breeds the Bikaneri, which is the strongest and the most useful for draught purposes, then there is the carefully bred fast racing camel the Jaisalmeri and lastly the Kachchhi, which is mainly a dairy animal. A one-humped camel is now being bred with long hair for cold resistance in N Iran and N Afghanistan.

Camels having spread in the normal manner – across land from North America across the Bering Straits and then through Asia, Arabia and Africa – very much later were taken by sea to two new destinations. Firstly back to their country of origin N America in the same way as horses had been taken before them. This was some years before the American Civil War, when the US Army imported and used camels to explore the South West desert region of the USA. When first mooted, the suggestion made by a couple of camel enthusiasts was met with the ridicule that innovations so frequently are. However after the usual bureaucratic delays lasting several years, finally in 1855 $30,000 was appropriated and this led to the birth of the US Camel Corps. After adventures with rogue dealers in Egypt and bribery, thirty-three camels were taken by sea to the USA. It was an amazing success with a survey being conducted of the Mexican Californian border through unexplored territory between El Paso and the Colorado River. The camels' performance was shown to be vastly superior to the horses and mules and not only through the desert. One doesn't generally think of camels swimming but this they did, across the strongly flowing Colorado River, which had already drowned several mules and horses.

A strangely contrasting and contradictory story is of a Colonel Earle having trouble in 1884 trying to cross the Nile with six hundred odd camels. These particular camels seemed very awkward in the water and quite a few drowned. With their long legs the camel is a little like the giraffe, the only mammal that cannot swim at all.

A more frivolous and earlier import was into Italy, when Ferdinand II (1620-1670) one of the de Medicis, who being quite an innovator and who had even supported Galileo and his theory of the earth revolving round the sun, kept camels as a kind of fashionable toy. Amazingly they survived right into the C19th. Another little venture was by a rather mad French adventurer called Jean de Bethencourt who pawned his Norman estates and with the help of King Henry III of Castile he set out to conquer the Canary Islands and then imported camels from Morocco with great success. They are still there and have now become a tourist attraction.

In Cairo until very recently the narrow winding streets, steps and blind corners were the result of not having wheeled vehicles. As recently as 1845 the width of a major new street being built was based on the combined width of two camels. Palanquins being carried between two camels were still seen in Cairo in the 1920s. Camels are now used very successfully on safari in Kenya and handbooks by English Kenyans give advice on training, harnessing, saddles and also general tips on planning your trip.

There is a rather awful story concerning an anti-pope, Gregory VIII who was consecrated at Rome in the presence of Henry V and ruled till 1121, but was taken captive by the Normans, mounted on a camel, facing its rear parts covered with dust and filth, then paraded before the real Pope Calixtus amid the insults and mockeries of the Roman mob, and consigned to a dungeon.

One question is why was the camel not used more as a draught animal in the early days? It was just as suitable as an ox if not more so in some areas.

The camel may have been the Beast of Burden across the deserts of Africa, the Middle East and Asia; it may have been the Gift of God to the Muslims and the mainstay of the Incense and Silk Routes but it is held in quite different regard in another continent. In Australia it is a pest, it is almost vermin. It all began when quantities of camels were imported between 1840 and 1907 to help in the construction of railways and the Overland Telegraph Line and the Goldfields Water Supply. These ranged from Bactrian through to Arabian riding camels. The first few arrived in 1860 for the ill-fated Bourke and Wills expedition, which was the crossing of Australia from Melbourne in the south to the Gulf of Carpentaria in the north, a distance of about 2,000 miles. Nineteen men took 26 camels, 23 horses and 6 wagons. Three of the wagons broke down almost immediately and amazingly they did not get round to using the camels as pack animals for a whole month. Bad leadership and planning led to most of the expedition leaving or dying.

Stud farms were started by Sir Thomas Elder at Beltana Station in South Australia. Very wisely Afghan cameleers were imported to care for and train the camels and were dubbed 'Ghans' for short. The camels were selectively and scientifically bred so as to fit them for the different tasks required of them and were ideal for using in the desert interior. Breeding there was so well selected that through the fifty years it operated the camels became superior to the imported ones. The numbers rose to 700 by the end of the 80's and by 1925 they had bred up to an astonishing 13,000. They were used to lay the Overland Telegraph Line and to help with the building of the cross country railway line. The longest North – South railway route in the world runs from Adelaide to Darwin a distance of about 1,850 miles and was called "Ghan" in their honour.

After motorised transport was introduced in the 1920's camels were phased out and released into the wild to become yet another not entirely desirable feral animal roaming the outback. There are now rumoured to be about 1 million. They have no natural predators and can live 40-50 years and start breeding at the age of 3-4 years. Their browsing and grazing are making evermore desert. Their search for water can make them foul up water places sacred to the Aborigines. They have also been known to break into boreholes and tanks and even bathrooms. Another drawback is that they can carry TB and brucellosis. What to do about them? Suggestions run from butchering them for their meat on an organised scale to letting the aboriginal youth have a go at hunting them to get them away from drink and drugs. But there is definitely a future for camels as milk producers and for their meat and their wool. The milk is known for its curative powers, and their meat is almost free of fat.

The other camelids live, as mentioned in South America. Until 25,000 to 30,000 years ago, there had been plentiful llama ancestors still living in North America. Charles Darwin, no less, was the first to find the skeleton of *macrauchenia,* a Pleistocene distant ancestor of the camel and the llama. It roamed in Argentina four million to 20,000 years ago. It was ten feet long and five feet at the shoulder. But the most extraordinary thing about it was its long prehensile trunk-like nose. It probably would have made a good pack animal if it hadn't unfortunately become extinct. An amazingly intact fossil skeleton of a camelid called *hemiauchenia* from the Miocene era (9 million years ago) was found in Florida. It is huge, over 7 feet at the shoulder and its head would have been about nine feet above ground.

The *camelid* branch that had moved south in the Americas evolved into two wild varieties – the guanaco and the vicuna; the Incas eventually bred llamas (*Lama glama*) from the guanacos and alpacas from the vicuna – the only exceptions to the Old World monopoly of domestic animals. Llamas in the High Andes were the equivalent of the Yaks in Tibet. Their drawback is their size and strength, they grow to 5 to 6 feet at the head and weigh between 120 and 200 Kilo but can only carry 25% – 30% of their body weight, which is proportionally more than a horse can carry but they are not strong enough to be ridden and if overloaded they simply lie down. The Spaniards coined the word "empacase" meaning 'to dig one's heels in'. Llamas are splendidly adapted to the high thorny *puna* and the rugged slope of the Cordillera, and can live 20 years, but they were not all that numerous in the Inca Kingdom of the Sun.

When Darwin visited South America on the Beagle voyage, there were vast herds of llamas roaming the pampas. He also described seeing herds of five hundred or more guanacos on the banks of St Cruz in1833.

Llamas also have the advantage, like yaks, of being adapted to the rarefied atmosphere of the high *puna* of the Andes and Cordillera. This is no mean feat as the Andes are the longest continuous mountain chain in the world and run through many climates. Temperatures can range from -30° to +35° in twenty-four hours. The Andean civilisations domesticated llamas in the region of Lake Titicaca possibly four or five thousand years ago, making it one of the earliest pack animals and down the centuries, they were carefully husbanded and selectively bred. Unlike their cousins the camel, Llamas have no hump, but a long thin neck with a small head and a cleft upper lip. This they can spit through when annoyed, even ejecting some of their stomach fluids, when very angry. Their feet consist of two toes and a tough pad and they are extremely sure-footed on all sorts of terrain from desert to high mountains. Like the camel and giraffe they pace, which means moving both legs on the same side and this leads to a swaying motion so perhaps it would have been unpleasant to ride them. In the wild they have about thirty in a group which consists of a single adult male and several females with young who stay with the group until about 12-15 months.

Although llamas do not store water like camels they can withstand thirst and they have the added advantage of being able to graze on absolutely any type of vegetation, facilitated by their three-chambered stomach. Their other main function was to provide wool; the coarser llama wool was for use by the common people and the finer fibres of alpaca and the vicuna wool were reserved for the nobility. Nor was that the end of their value, as they could also be eaten and used for sacrifice. They were classified as agricultural resources and typically divided into three groups; the first was for the priests and their various religious ceremonies; the second was for government use; and the third was for the general public and general community work.

Being the only beasts of burden for the Incas they were absolutely indispensable for their military campaigns and for moving the vast treasures of the empire. Annually long caravans took salt for trading for maize and sugar. Then the Conquistadors led by Pizarro arrived in 1531. From 1545 to 1640 silver was taken from the Potosi mines in Peru, which are situated deep in the cordilleras of the Andes at 12 – 17,000 ft above sea level and 400 miles from the sea. Llamas proved to be indispensable for carrying this silver and Simon Bolivar, the conquistador, estimated that they were using 300,000 in those mines alone. Every year a great ceremonial hunt was organised by the Inca kings called a 'chacu' in which 20,000 to 30,000 formed a circle to trap the animals.

The pastoral way of life created close bonds between the herdswomen and the llamas, as in some parts there was the same total reliance on the llamas as the Tibetans have on their yaks. Herders had enormous vocabularies to describe alpacas, as many as nineteen different colours and tones; then there were twenty-six names for patterns for the two sexes and these could lead to 20,000 different description combinations of words possible.

Meeting up with the strange new animals, the llamas, there was a difficult reaction from the Spanish whose ideas of natural history were set in concrete based on the ancient system and Pliny's sacrosanct authority. However a Spaniard, Fernandez de Oviedo (1478 – 1557) who was really the first field researcher of modern natural history, wrote the first comprehensive history of Spanish America, the history of the Spanish discovery, conquest, and colonization of the Americas from 1492 to 1547, along with descriptions of the land's flora, fauna, and indigenous peoples. His *Historia*, which grew to an astounding fifty volumes, includes numerous interviews with the Spanish and indigenous leaders who were literally making history, also the first extensive field drawings of America rendered by a European, then there are reports of exotic creatures, ethnographic descriptions of indigenous groups, and detailed reports about the conquest and colonization process. Orviedo found the llama the most useful animal he had come across in the whole of South America and judged it to have been 'created by God to sustain his children'.

José de Acosta (1539-1600) was called the Pliny of the New World and deserves a great deal more recognition. In his book "The Natural and Moral History of the Indies" he gave the very first detailed accounts of the geography of the area and the natural history. He also has written up the Aztec and Inca histories, and even tells of the use of the coca leaf. This latter was what gave men the strength to carry up to their own weight at the high altitudes of the Andes. He even tried to explain in biblical terms why there were *camelids* in South America – and even, most amazingly, recognized there must have been a land bridge connecting to Asia. He even suspected the existence of Australia. He was a pioneer of geophysical sciences and right ahead of his time.

Philip II of Spain (1527-1598) ordered the Audiencia in Lima to send 140 female and 60 male llamas to breed in Spain. He also ordered vicunas sent but unfortunately they couldn't adapt. But still the orders to stop killing llamas were largely ignored. The early settlers nearly hunted them to extinction as instead of using the local animals, they, imbedded in prejudice, tried to introduce sheep and goats. Later Mules, introduced by the thousand in the eighteenth century, largely supplanted the Llamas. Now they are being bred back and in the United States alone there are about 100,000. They have quite a cachet in England too.

Like the Arabs with their camels, the Indians decorate their llamas and alpacas with bright tassels and packs. Over 65,000 rural families totally rely on these animals and pursue a family tradition of selective breeding. They also export 4,000 tons of alpaca fibre every year, and the women practice weaving and rug making. They call the llamas their 'silent brothers'. A herdswoman particularly has very close emotional links to her llamas and believes she should look after her herd as if they were her children. In fact they believe that llamas have souls, feelings, intelligence and even possess a memory. The guanaco's wool is double coated with coarser guard hairs and soft under coat, and rivals cashmere and pashmina from Asia. They are never used for milk. A pure colour is much prized but most are variegated black, brown and white and some are covered with blotches of every shade giving a clown-like effect. The wool of the alpaca is highly prized as its fleece is amazingly warm and soft and having no lanolin it can be worn by allergic people. If the alpaca is not shorn regularly its coat can reach the ground. To rid itself of parasites and lice it will roll in the dust. Unfortunately the guanaco and the vicuna are heavily hunted in spite of efforts at conservation.

ET CETERA

1. And now for a recipe for cooking camel: In a novel by the Czech author Bohumil Hramal there is a most charming two-page episode on preparing a feast for the Emperor of Abyssinia. The main dish was Stuffed Camel. The preparation was elaborate:

first, the cooks the Emperor had brought with him to Prague boiled several hundred eggs. Then they stuffed twenty turkeys with white bread-based stuffing, and baked them. These turkeys were then used as a stuffing for two antelopes, which became in their turn the stuffing for the camel. In addition fish were added to the camel's belly, the remaining boiled eggs were used as padding, and a strong seasoning was generously administered. The camel was then grilled in the courtyard of the hotel. This amazing dish fed over 300 people."

2. *"The 'orse 'e knows above a bit, the bullock' but a fool,*
   The elephant's a gentleman; the battery-mule's a mule;
   But the commissariat cam-u-el, when all is said an' done,
   'E's a devil an' a ostrich an' a orphan-child in one."

   Rudyard Kipling 1865-1936

3. Aristotle believed that the giraffe was a cross between a camel and a panther.

4. An innovation in the United States is to train a llama as a golf caddy. But it must obviously carry two bags of golf clubs for balance.

5. "The Camel's Hump is an ugly hump
   Which well you may see at the zoo;
   But uglier yet is the hump we get
   From having too little to do".
   Kipling 'How the Camel Got his Hump'

6. "The Seven Pillars of Wisdom" written by T.E.Lawrence in 1926 was full of instructions about dealing with camels, but he also rode around a great deal in a Rolls Royce.

7. Geoglyphs in the plains of Peru 100 miles south of Lima the Nazcas Lines from 100BC – 700AD were made by removing the desert varnished rocks from light coloured soil. There are huge depictions of llamas and even a human.

8. Allah has 100 names according to the Arabs and the Koran contains 99 of them. Only the camel knows the 100th name, which is why he looks so superior.

---ooO0Ooo---

CHAPTER 6

TRADING & TREKKING

Trading and Trekking ...carriage

Oxwagon

(Photo: South African Railways & Harbours Publicity & Travel Dept.)

"The untenanted Kosmos my abode;
I pass, a wilful stranger;
My mistress still the open road,
And the bright eyes of danger."
R.L. Stevenson (1850-1894) "Youth and Love"

We jump in our cars, we catch a train or a plane and in no time we are at the shops, the other end of the country, or the other end of the world and think nothing of it. But these forms of transport are all very, very recent. Steam engines were evolved by James Watt in 1765. The bicycle, especially liberating for women, was only produced in the 1860's. The ubiquitous car was invented and adapted by a range of people from about 1889 mainly in Germany by Daimler, and then they were produced in quantity in the USA by Olds (the Oldsmobile) in 1901 and the famous Model T Ford came off the production line continuously from 1908 to 1941. Before these inventions, we were entirely dependent on animals for transport and that is not long ago.

Trade routes existed long before the Romans invented roads. Central Asia and Eastern Europe have been linked for at least seven thousand years by prehistoric paths and trails, across deserts and mountain passes.

These were the strands that became woven into the longer trade routes. Their very names indicate these were mainly for the purpose of trading, The Silk Road, the Spice Route, the Incense Road and the Fur Road. They really came into their own after the Camel, the Yak and the Horse had been domesticated. Religions such as Buddhism and Zoroastrianism, Mannicheism, Christianity and Islam also travelled along these paths, as so unfortunately did warfare. Tools and new crafts and techniques also were passed along, the building blocks of education, of civilisation. Later still travelled the explorers and archaeologists. This is a gigantic subject and I can only include a précis of what could easily be a book on its own, but it is vital to my argument to mention the trade routes.

One of the most famous routes became known collectively as "The Silk Road". The geographer Baron Ferdinand von Richthofen gave it this romantic but slightly misleading soubriquet only in 1877. These ways, dating back to at least 200BC crossed some of the most inhospitable territory in the world, mountains such as the Pamirs and the Hindu Kush and the Karakorams; deserts such as the Lop, the Gobi and the Taklamakan. Towns with equally evocative names like Samarkand, Kashgar, Dunhuang, and Bokhara provided depots, oases and caravanserais. The Pamirs soar up to 7,000m, and their name in Farsi means 'The Roof of the World'. They are connected to the Karakoram, the Himalayas to the South, and the Hindu Kush to the West. The inhabitants – the Pamirans live in the valleys which themselves are at 2,000m and practice another branch of Islam called Ishmailism, a form of Shiism. On the icy plateaux the ice can be as thick as 150 metres.

The discovery in 1991 on an Austrian mountain pass of Őtzi, the corpse of a Neolithic man, well-dressed and well-armed, dating from 3,300BC, showed that these passes were in use at least as long ago as that. Baltic amber has been found in Greece and Mycenae, also stones for axe heads, obsidian and dolerite flint have been located miles away from where they were mined. Lapis lazuli from Afghanistan was found in the pre-dynastic Egyptian site of Naqada (3300 – 3100). All these examples show that trade was taking place thousands of years ago, over vast distances.

Most maps show the Chinese departure point as Xi'an, the old capital of the Han dynasty (the home of the famous Terra cotta Army). Very few people travelled the full length of these ways, which stretched 7,500 miles from end to end. Goods would change hands perhaps many times before reaching their ultimate destination. Although horses and donkeys were used to some extent the Camel was really the main mode of transport. Camel caravans will always be associated with the Silk Road. Indeed without them and their ability to withstand lack of water, extremes of climate, the terrible terrain, and their capability of carrying 700lbs to 800lbs, the merchandise could not have been moved the great distances in the conditions that existed.

> *We travel not for trafficking alone:*
> *By hotter winds our fiery hearts are fanned:*
> *For lust of knowing what should not be known*
> *We make the Golden Journey to Samarkand.*
> (Hassan by James Elroy Flecker)

What was the going like? Well roads built for purpose did not exist until Roman times. So the Silk Road really consisted of a complex network of well-worn tracks and paths, merging and parting. From East to West it took at least six months to traverse and from north to south about two months. After leaving China the route led along the Great Wall and the Gansu Corridor before splitting at Dunhuang into two Northern Routes and one Southern Route. (Dunhuang is where the best-preserved examples of Buddhist cave art is found in the Mogoa Caves.) These routes avoided crossing, as far as possible, the great deserts of Gobi in the east, and Lop and Taklamakan more in the west. The longer of the Northern routes was less arduous than the Southern. It ran between the Taklamakan Desert and the towering mountains of Tian Shan (Celestial Mountains) and passed through various oasis towns, notably Hami, Turpan, Kucha and Kashgar. From the highest to the lowest is Turpan with its vineyards and nut trees, which lies in the third deepest depression in the world at 154m below sea level. The Southern route ran through such oases as Miran, Charklik, Cherchen, Khotan and on to the junction of Kashgar. At Kashgar there were several ways to take, one was westwards to Europe via the famous cities of Tashkent, Samarkand, Bokhara and Merv; others led south via Balkh over the Hindu Kush to Kabul and Peshawar. Fed by melting snow led through underground channels from the mountains, these oases had trees and fruit, pomegranates and grapes. To the arriving traveller these fruit orchards must have seemed like glimpses of Paradise.

The caravanserais offered many services apart from water, food, lodging and rest. Goods were dropped off and taken on. Traders mostly travelled only some of the way. They also provided banking services, blacksmiths, and a change of horses or camels. Most of the water for these desert towns came from an enormous underground irrigation system fed by the glaciers of Tian Shan.

No doubt people did improve the routes by moving rocks that blocked a mountain pass to make it less dangerous, or filling in deep ruts with stones or soil; or varying their route in certain conditions. They had to cross arid deserts where the temperatures could vary from scorching days reaching +38° in the brief summer to freezing winter nights of −40° in the 8 month long winter. A missionary monk, William of Rubruck on a papal mission in the mid C13[th] reported that the intense cold, even in May, froze his bare feet. Travellers had to contend with dust storms, lack of water and even hallucinations brought on by atmospheric conditions and no doubt stress and exhaustion. Indeed in the Gobi desert, which they traversed at night, travellers often experienced a strange dancing magnetic light that looked like distant campfires. Marco Polo described strange spirits that like sirens tempted you to leave the track and follow them. The Lop Nor originally held a lake into which the Tarim River ran, but it is now dry and consists of just yellow grey clay terraces. Even most of the sand has blown away in sandstorms. The Taklamakam Desert or "Sea of Death" is the largest sand only desert in the world and has giant sand dunes 100 to 300 metres high, and these dunes travel by wind power 150 metres a year and naturally threaten to overwhelm the oases. 'Taklamakan' roughly translated from the Uighur tongue means, rather grimly, "If you go in, you won't come out". Sand storms are one of the greatest hazards in the sand deserts. Old experienced camels seemed to sense their arrival and would bury their noses in the sand. Fortunately they were furnished with their double rows of eyelashes that gave them protection. The drivers would take the hint from the prescient camels and cover their faces in felt. Luckily the sand storms usually passed over quickly.

And what did they use for transport? In about 3,500BC, the greatest innovation in the development of transport − the Wheel was invented.

However the camel completely ousted wheeled transport in Muslim and Near East countries. For several hundred years from about the C3rd right into the C20th, the pack Camel, and to a lesser extent the donkey and mule took over. The use of the wheel was not dropped for any ideological reason as the water wheel and the potter's wheel continued in use. The reason was that a loaded camel was found to be far more economical and suitable all round than mules, donkeys and oxen hauling wheeled vehicles. Oxen are extremely slow and only cover about 6 – 9 miles a day, whereas a camel can manage 20 – 25 miles. The ancient wagons were heavy and cumbersome and more likely to break down and deserts and high mountains are, of course, very short of wood for making or repairing carts. Also a wagon needs two animals and a man to drive it whereas one man can control a train of six camels on his own. The wheel for transport was marginalized even in non-Arab countries such as Iran and North Africa, where the camel had been introduced in Roman times. India was one exception and carried on with carts.

The camel is totally suited for the inhospitable terrains it had to cross; its thickly padded feet are ideal in sand or stony ground and do not require shoeing; in fact it not only does not require paved roads, it actually finds the unpaved ones more comfortable for its feet, and no shoeing is required. Then other advantages were that it needed far less food and water than a mule and could carry twice as much weight; also they live much longer and the initial outlay is far less. They could cover about 40 miles a day, so on all points the camel scored. And according to a report made, centuries later in 1894 by Major Leonard, the British transport officer, the camel was still the main carrier on these rough hewn routes. But after the C15th the central Asian caravan routes did start to go into a decline.

The only requirements for the caravans were bridges and some sort of accommodation every twenty odd miles – a village, an oasis or a caravanserai. Another problem was that in the C2nd the Nabateans of Palmyra, a central oasis of Syria, taxed donkey carts four times as much as camels in order to protect their home-grown camel caravans. The Nabateans had trade routes to transport about 3,000 tons of myrrh and frankincense a year from southern Arabia and as sailing north up the gulf to Aqaba is made almost impossible by the opposing wind and currents; they used their caravans to bring the incenses to Damascus, Gaza and Alexandria.

And what did they trade in? The list of goods is endless. Curiously, one of the first was Jade from the Khotan area on the Southern Silk Road. It predated silk in importance by 5,000 years and was highly prized by the Neolithic people, in spite of it being incredibly hard to work. Salt was another very early commodity. Silk obviously was the most famous export from China, reaching the Mediterranean about C2ndBC. The Romans acquired a huge hunger for it. Threads are spun from the cocoons of a special moth (*Bomybx Mori*). For five weeks before their metamorphosis the caterpillars are fed on mulberry leaves. Legend has it that Xi Ling, the wife of the Yellow Emperor (2698 – 2598BC) first created the magical fabric. The Chinese managed to keep the creation and therefore the monopoly of silk a closely guarded secret for centuries. People in the West actually thought that silk was combed from the leaves of trees. By the time of the Han dynasty (206BC –220AD) production had become very sophisticated with silk being woven in every colour and variety, even figured and painted. Another legend has it that a second Chinese princess arrived in Khotan in 440AD having smuggled cocoons out in her piled up hair. Another tale has it that Nestorian monks hid silk worms in their walking sticks. Whatever way the secret got out and after that silk was no longer a monopoly of the Chinese. However they still remained the main producers. Khotan also had a thriving paper industry, and rugs and wool.

Other luxury goods carried along the Silk Roads were gold, ivory, Lapis lazuli, rubies, diamonds and pearls. Amber from the Baltic was another exotic import, and coral from the Mediterranean and furs from the Russian steppes. Glass was another import, as it was not manufactured in China until the C15th. An important innovation was the Chair which arrived in about the C2ndAD. Spices became fashionable, and more than necessary to disguise the flavour of meat going off in the days before refrigeration. Rhubarb was a crop highly valued as a purgative. Other plants included saffron, narcissus, pistachios, peppers and asafoetida. The last was to conceal foul odours, of which there were plenty in the days before baths, refrigeration and general hygiene. Other vegetables were introduced such as spinach from Persia, also kohlrabi – called the Corinthian turnip by Pliny, and sugar beet. Indigo came from Persia and was used not only as a dye, but as a cosmetic. Then there was a poison called cinnabar which contains mercury and was also valued for its red colour, often employed in lacquer ware. A curious animal medicine was 'bezoar', partially digested food or hairballs often found in the fourth stomach of ruminants and thought to be an antidote for poisons.

Animals too were taken along the Silk Road, of which horses were the most prominent. The game of polo from Persia caught on in a big way during the Tang dynasty and was even played by women. The beautiful Ferghana horses were always being imported for breeding stock. Then there was a fashion for dancing horses, which were decorated with wings and horns and precious metals. Hawks and falcons were hugely popular for falconry and surprisingly red-necked ostriches from Syria (now, of course extinct). Certain dogs were sought after, ranging from hunting hounds to lap dogs. Yaks from Tibet went to China as a tribute and so did certain curious breeds of sheep from Samarkand and the Pamirs.

Less concrete but even more important were the ideas carried and interchanged along the route, which included religions such as Zoroastrianism, Buddhism, Christianity and Islam, new technologies and political ideas.

The Silk Route is not all uninhabitable deserts. There were and still are a great variety of wild animals. Two of the most famous are the Przhevalsky horse, now no longer in the wild but only in zoos, and the huge Marco Polo sheep or *argalis (Ovis poli)* owning the most amazingly huge twisting horns but also threatened in the wild. (The modern day explorer, Ralph Cobbold shot one with a horn span of 64".) There were also ibex, wapiti and gazelles, both red and brown bears, snow leopards (*Uncia uncia*), wolves and wild boars. Smaller mammals were foxes, marmots and the Tibetan hare. There was bird life too – pheasants, woodcock and duck for the pot. Ibises, owls, plovers and larks abounded in the right places. Where there was water you would have found ospreys, grebes and herons.

The Routes passed through an array of nationalities and tribes, who varied in costume and custom, language and physiognomy. Greece and Rome were one end and the Chinese (certainly not homogeneous either), the other. One of the most important people were the little known Sogdians of the area known as Transoxiana (Eastern Iran), who centred on Samarkand and Bukhara. Historically they had existed from C6thBC – C10thAD after when they vanished into the Persian Achaeminid world. However they dominated much of the trade into China. They also introduced

Lucerne and vines, which were used to feed the Heavenly Horses that were bred from their neighbours, the Ferghanas. Originally the Sogdians had been a very warlike people but Alexander the Great had devastated and subdued them in 329BC. (He married Roxane, the daughter of one of their noblemen.) After that they were transformed from warriors into merchants. Their language Aramaic became the *lingua franca* of the Silk Road. In the C5th and C6th many of the Sogdians that had immigrated into China bore the title "Sabao" derived from an Indian word meaning chief caravanner. For themselves they first imported Chinese paper and then began making it themselves in Samarkand.

The bazaars in the various towns were and still are thronged with a huge variety of people, in different tribal dress, with varying physiognomies, religions and customs. The Uighur women wear lovely striped silk and the men have embroidered black velvet skullcaps and black velvet coats with vivid linings. But the Uighurs were not just a picturesque tribal people for they dominated the Silk Road for a thousand years, right up until the Manchu invasion in 1759. It was the communists who dealt the final blow to their culture. In the early C20th and even before, archaeologists unearthed amazing relics of their civilisation, which are now displayed in major museums in London, Tokyo, Paris and other capital cities. They had great literature for which they used a variety of scripts such as Cyrillic and Latin. The Sogdian italic script was used for 800 yrs but after embracing Islam in the C10th they adopted the Arabic alphabet, and its use became common by the C11th. They also had an extensive knowledge of medicine. Some Western scholars attribute acupuncture to them and not to the Chinese. Chinese envoys were greatly impressed by their architecture, music, painting and sculpture. They had mastered the art of printing long before the Europeans.

Also along these routes went explorers and travellers. Their discoveries and writings opened up whole horizons to the rest of the world who before the advent of modern travel and modern communications had no idea of how other nations lived. Without the camels and to a lesser extent horses these travellers would not have been able to cover the distances over the harsh terrain that they did.

One of the earliest and most famous travellers along the Silk Route was a Buddhist Monk named Xuanzang (602 – 654AD). He took the

Northern Route from China via Tashkent and Samarkand, through Kashmir, round India and then home back to China along the Southern Silk Route. The 10,000mile journey took him twenty years from 629AD to 649AD. Apart from deserts, he had to cross three of the highest mountain ranges of Central Asia. His writings give us a unique glimpse of the Silk Road in a particular time of its development. They also inspired Aurel Stein who followed the routes eighteen hundred years later. Xuanzang was particularly intrepid as not only was he travelling in the harsher conditions of those days, but to embark on the adventure at all, he had had to smuggle himself out of China in the first place. The reason for this was an edict from the Tang Emperor forbidding all travel to the West. However he managed to pass through the Jade Gate and set off across the Gobi desert with a horse and guide. But his journey nearly finished there as first his guide tried to murder him and then he dropped his precious water.

Undoubtedly the most famous Westerner who travelled the Silk Road was Marco Polo (c.1254 – c.1324). It all began when his uncle and father Matteo and Niccolo Polo who were trading round the Black Sea suddenly elected to go east and then found themselves marooned in Bukhara for three years. Unexpectedly rescued by the Mughal ambassador, they were persuaded to accompany him to meet Kublai Khan (1214 – 1294AD). And so they travelled along the Samarkand, Kashgar, Gobi desert route then along the Hexi corridor and finally reached Beijing, the new well-built capital. Kublai Khan, as promised, was delighted with them and after a year they set off home armed with a sort of VIP passport – a gold tablet one foot long by three inches, and inscribed with an admonition to assist the bearer for the Great Khan or be killed. The gold tablet meant that horses, inns, food and guides were provided along their route, which was just as well as it took them three years to return home.

They then set off for Cathay, as China was then known, for the second time taking 17 year old Marco with them, and letters and gifts from the Pope (Gregory X) for Kublai Khan. They considered going by sea but the boats were of terrifyingly poor quality, stitched together with coconut fibres! So they took another variation of the Silk Road through Armenia, Afghanistan and over the Pamirs. This was the first time the name appeared in history. The Gobi desert impressed the young Marco with its size and its impenetrability. He reckoned it could take a year to travel from end to end and a month to traverse.

Marco Polo was very observant and noticed that while the women did most of the work the men spent their time in hunting, fighting and falconry. A not unusual division of labour! He also noted "They drink mare's milk subjected to a process that makes it like white wine and very good to drink. It is called koumiss."

It was now three and a half years and five and a half thousand miles since they had left Venice. So much must have been new to him but, amongst other things, he remarked on asbestos, paper currency, coal and The Imperial Post. This last was an amazingly efficient means of communication, organized in a three-tier system. The lowest grade of dispatches was carried by foot-runners with relay stations 3 miles apart. The runners wore bells to ready the relief thus avoiding wasting time in handing over. This system meant that a message took only 24 hours to cover a distance that normally took 10 days. More urgent mail was taken on horseback and the relay stations were 25 miles apart. But the top-priority dispatches were conveyed by a relay system. Instead of bells the dispatch rider would sound a horn and a fresh horse would be awaiting him. According to Marco Polo these couriers could travel up to 300 miles in a day.

But the big question is did he really go to China? And live there for 17 years? His book was known in his time, as "The Description of the World "or "The Travels of Marco Polo" and it became the greatest travelogue and bestseller. It is full of amazing information but yet there were some extraordinary omissions. For example he never ever mentioned such noteworthy habits as tea drinking, foot binding or calligraphy. Even in his time the book got dubbed as "Il Milione" or The Million Lies" But he swore he hadn't 'told the half'. And, of course, such was the ignorance of the time that his fabulous tales of the exotic customs of the East must have been hard to swallow. But either way the Silk Roads connections with China and the Polos' travels had opened up a view of the Far East and its inhabitants far more civilised and impressive than the myopic Europeans had ever imagined. Even 200 years later famous explorers and adventurers such as Prince Henry the Navigator (1428) and Columbus (1492) carried Marco's book with them.

Another great but little known traveller was Ibn Battuta, born in Tangier, Morocco in 1304. He took to travelling when he was twenty-one years old and continued for the next thirty years, when he returned to Fez and dictated the accounts of his journeys. This book is known as 'The Travelogue (Rihla) of Ibn Battuta'. He visited every single Muslim country of his time and ranged as far afield as Mombasa and Timbuctu in Africa. But, under the patronage of Sultan Mohammad Tughlaq he also travelled the Silk Road, visiting important cities such as Bukhara, Balkh, Herat, Tus, Mashhad, Tabriz and Nishapur; he crossed the Hindukush Mountains via the 13,000 ft Khawak Pass into Afghanistan and passing through Ghani and Kabul entered India.

Ibn is supposed to have covered 75,000 miles, making him one of the most remarkable travellers of all time especially when you consider the modes of transport available to him in his day. The big question about him is not whether he went to the Far East but why is it not generally known? Neither in the western world is much information available, but more mysteriously not in the Muslim world either.

None of the modern explorers ever got married. They seemed to be drawn to isolation and privation. The earliest of them was a Russian Nikolai Przhevalsky (1839-1888) the man after whom the famous ancient breed of horses is named. He was sent into central Asia by the Imperial Geographical Society to map parts of central Asia. He surveyed 7,000 miles and went on three expeditions as far as the Ordos plateau in North China and to the edge of Tibet. His attempts to reach Lhasa were continually frustrated even though he was carrying 70lbs of Turkish delight as a bribe! He found the British and Indians had been there first and made maps. However he did succeed in bringing back an amazing array of animal and botanical specimens. There were delphiniums, rhododendrons and a range of honeysuckles but the most important of course was the primitive wild horse sub specie.

Another interesting but wild adventurer along the Silk Road was Sven Hedin (1865-1952), a Swede, trained in physics, geology and zoology at the University of Stockholm. He was almost pathologically anti-social, but physically reckless. He made four expeditions to Central Asia between the years 1893 and 1935 mapping the Pamir Mountains and the dreaded desert of Taklamakan. He also visited Tibet and tried in vain to get to

Lhasa, where he was pre-empted by the British in the shape of Sir Francis Younghusband. Owing to his pro-Nazi attitude, Hedin was stripped of all his British honours, including a knighthood. He was on the Russian side in the Great Game – the name given to the Anglo-Russian rivalry and wars mainly involving Afghanistan. Owing to his foolhardiness, he nearly died more than once and camels, servants, horses and dogs were sacrificed along the way. However he did find the site of Loulan, lost for hundreds of years in the Lop desert, and filled in quite a few other gaps in the map. He reported that the Tibetan horses were fed on meat. But thanks to him the Chinese were able to motorize part of the Silk Road and so keep control of Sinkiang He wrote three major books, *The Flight of Big Horse* (1936), *The Silk Road* (1936) and *The Wandering Lake* (1940).

Sir Aurel Stein (1862-1943) was a Hungarian Jew and in strict contrast to Hedin he took the greatest care of his men and beasts. He travelled by every means, on foot and by yak, camel or pony. He had been fascinated by the Silk Road and the travels of Xuanzang as a child. He studied Sanskrit and Philology at the university of Vienna and Leipzig. Then he moved to England and did research in the Ashmolean and British Museums and the Bodleian and India Office libraries. His ambition was to follow in the footsteps of his hero Xuanzang, and his aim was to find traces of lost civilizations. He led three major expeditions over seven years and travelled 25,000 miles. Lord Curzon, the Viceroy sponsored his very first expedition. He was well equipped with two cooks, a surveyor and his first black and white terrier named Dash, which turned out to be the first of seven terriers all called Dash. In the hostile Taklamakan desert he found fragments of frescoes, scripts in long dead languages and other archaeological treasures including a painting on wood of the famous Rat King, which saved Khotan from the Huns by chewing through their bridles! He also uncovered Buddhist paintings, sculptures and texts. It was unbelievably cold, in spite of furs and heaters and Dash sharing his bed. As an extra he had to deal with a fracas between one of the cooks and his surveyor and separate them by force.

In 1907 Stein went on his most famous and most productive expedition to the sites of Loulan, an ancient oasis city dating from C2nd BC on the edge of the Lop desert. His greatest discoveries were the "Caves of the Thousand Buddhas" at Dunhuang, in NW Gansu province. The self-appointed guardian of the caves was the Buddhist Abbot Wang. He bribed Wang to open a cave and there found a solid mass of manuscripts ten feet high, written in Chinese, Sanskrit, Sogdian, Uighur and Tibetan. Probably the most important was 'The Diamond Sutra' dated 863AD, the oldest known printed document. In all, he took nearly thirty cases of manuscripts and other relics for the British Museum. Everything was incredibly well preserved by the harsh dry climate.

He crossed the Taklamakan desert yet again, from North to South, relying entirely on his compass, as there are absolutely no landmarks in this featureless desert. Water was carried in the form of ice blocks and was just running out when they found the Keriya River. Stein had a bath in his canvas bath to celebrate. From Khotan he sent home fifty camels laden with archaeological treasure. At the burial grounds of Astana he found dried up mummies buried with dumplings and cakes over a thousand years old.

People may argue for or against the rights of taking treasures from another country but at the very least they have been preserved for posterity and investigated by scholars. When the Chinese became aware of the drain of works of art they intervened and Stein's last expedition was dogged by government interference. As it turned out the Chinese were not in the least responsible about managing the excavations – works of art and manuscripts turned up for sale all over the place in the open markets.

Stein was responsible for putting the Silk Road back on the map. Owing to the improved navigation on the Indian Ocean, the trade routes had become more sea going and the Silk Road had all but died. Luckily Stein's meticulous research, annotation and photographs produced an invaluable historic legacy of one of the most significant trade routes of the world. His best-known book is *On Ancient Central-Asian Tracks*.

The particular commodities the Russians merchants' traded in were cotton more than silk and then tea by the ton for their samovars. By the early C18th it has been estimated they were receiving 600 camel loads of tea annually. At first it was prohibitively expensive and only for aristocrats but by the end of the century ten times the amount was coming in and samovars were humming in many more households. The samovar itself is a converted Mongolian cooking pot, the idea imported together with the tea. By the mid C19th no fewer than 10,000 camels were employed in the tea trade.

Camels were used, for many years on the Hajj, the pilgrimage to Mecca. Caravans of 1,000 to 2,000 camels were used. Kings built hotels at the caravanserais to encourage all this trade and pilgrimage.

Horses were actually used long before camels arrived to cross the Sahara on various trading journeys. Another enormously important and historic camel route, dating from the Middle Ages, ran from Timbuktu to the salt mines of Taudenni in Mali which lies about 800km (500 miles) to the north of Timbuktu. Salt was and is a very sought after and precious substance. Timbuktu became not just a centre for a rich trade but also of Islamic study. And its very name became a word for mystery and inaccessibility. There was no specific and easy route to follow and the journey could easily take two weeks. To this day the caravans risk crossing this deadly part of the world where the least mistake in direction can lead to certain death. And indeed in 1805 over 2,000 people and 1,800 camels died when they missed the route from Timbuktu to Taudeni. The Tamashek guides go by wind patterns in the sand, by the stars and by a sixth, almost mystical sense of direction. But now the Islamic wars and the use of 4x4s are threatening the whole sacred ritual. Precious manuscripts have been destroyed and salt prices have dropped.

Leptis Magna owed its wealth mainly to the camel trade. The Eastern caravan route was one of the oldest ways in the world, called *darb el arbaïn*, which meant "The Road of 40 Days" and was the link between the Nile and El Fasher in the Sudan. They brought goods from further south in Africa such as ivory, ostrich feathers and gum Arabic. But the most important imports were Gold and Slaves from West Africa. These were exchanged for arms, cloth and trinkets. Ghana in the C11th was so rich that even the dogs had gold collars and the king was dubbed 'King of Gold'. In the C14th the King of Mali could easily afford to pay 10 metric tons of gold to visit Mecca. The slave trade took place in appalling conditions as the captives had to walk across the Sahara in chains and many died on their way to North Africa to become servants and concubines and eunuchs. In 1848 the trade was abolished by the French but it still went on under cover and is probably still going on, in a small way. Salt also became part of the slave trade and a foot sized lump of salt was worth one slave. Returning to history, traction power was all important in the exploration and colonising of North America and southern Africa. Indeed the Europeans, who rode horses, wore armour and wielded steel weaponry and firearms had an unassailable advantage in the original conquest in the C15[th] of the Mayans and Aztecs and the Incas.

We had chariots and carts and sleighs and carriages. What we didn't have were roads, until the ingenuous Romans realised they were essential for the growth and maintenance of the Roman Empire. Called 'viae' which means to go with a vehicle, they constructed between 50,000 to 60,000 miles of roads which ran all over the expanding empire. One of the oldest and still in use is the Via Appia dating from around 312BC. The best roads consisted of four layers about a yard in depth with rubble at the bottom for drainage and finished by flat stones for paving. The Laws of the Twelve Tables drawn up c.450BC, specified such things as the width (8' to 16') and the right to cross private land. In urban areas vehicles were forbidden except for certain ladies and officials. Romans rediscovered concrete which had been used by the Egyptians, for the paving stones and pioneered its use for bridges. Milestones measured the roads, and were erected every thousand paces or 'milia pasuum', from which the word 'mile' is derived. The saying was 'All roads lead to Rome' and in Rome the 'Golden Milestone' was set up to act as the centre from which all roads that led from Rome were measured, much like Hyde Park Corner is used as the central destination in London today.

Drawn by oxen, mules or horses there were three main conveyances, two-wheeled carts, carriages for groups and larger carts for freight. At convenient intervals along the ways were inns some for officials only and others for the general public. The frequently travelling armies set up their own camps. As well there were two postal services one public and the other private. For urgent business relays of horses and riders were used and in extremis a letter could travel 500 miles in 24 hours. But wheels need roads and when the Roman roads fell into decline so did the use of the wheel.

Away from Roman influence, across the rest of the world there were mostly just tracks, and customary routes to follow. However here and there around the world exists a few remnants of pre-historic roads. Most have vanished without trace; many are overlaid by modern roads. However from time to time archaeologists do find remnants of ancient ways. People crossing frozen and snow-covered lands obviously left little trace. Tracks through the jungle were only temporary and soon overgrown. But, according to the terrain, a few roads or prehistoric ways have been preserved.

Moving goods overland was far more expensive than by sea in the time of Diocletian's Rome. For example wheat from Alexandria being sent 1,250 miles by sea was actually cheaper than 50miles by land. Practically every journey, either overland or by sea, was hazardous and strenuous and it took years to travel from Europe to the Far East. This enormous increase in trade and camel breeding led to an economic and unfortunately a military transformation.

Then coming down to the C18[th], in order to cross the Great Plains and the Rockies in North America, the pioneers used wagons drawn by teams of oxen, mules and sometimes horses. Many of these pioneers were immigrant Germans, Irish, Scots, or English who had been farming in Pennsylvania, Maryland and Virginia. The wagons were known by several names – Covered Wagons or Conestoga Wagons (named after the town where they were made). They were also nicknamed 'Prairie Schooners' partly because they looked a bit like a ship sailing over the tall grass, which hid their wheels and partly because they were actually boat shaped complete with rowlocks so they could be floated over a river while the oxen or other draught animals swam across. One special characteristic was that they tilted up slightly at both ends to minimise the sliding away of equipment when the wagon was going up or down hills. The smaller lighter versions were the ones mainly used by the new settlers driven out by over crowding and moving west to take up the 'empty' land that was to become, in time, Oregon and California. The new settlers took the Great Wagon Road to North Carolina and on to Georgia. At one stage the Wagon Road ran along the Shenandoah Valley following the former Great Warriors Road of the Amerindians and right on down to the Savannah valley. Later a heavier variety of wagon was developed that could carry a considerable weight, even up to seven tons or more and was extremely strongly built and so could carry freight over the Appalachians.

In 1835 came the South African Great Trek, partly a sort of monumental tax evasion of the Dutch from the British. But there were more reasons than that. They reckoned that the British in their domination were breaking down the old social order of white supremacy over black, which they saw as "God's Will". And so they migrated looking for a better life over the horizon as people have done before and are still doing now. Huge spans of oxen dragged the Voortrekkers' wagons across the mountains away from the Cape. A span usually consists of fourteen oxen but would be increased according to the load and the terrain. Each ox knew exactly where its place was along the trace and the name of its position. The leaders were the most intelligent and could respond to simple commands such as 'Hot!' (Left) and 'Haar!' (Right). Where the going got too steep the wagons would be taken apart and loaded onto the oxen and reassembled when the route made it possible again. Going down steep mountainsides presented another hazard and the trekkers solved this in novel ways, dragging a tree or even removing the back wheels all together and lashing big branches that protected the axle and acted as a primitive kind of brake/sledge. Oxen may look tough but they are actually not as strong as they look and on the trek had to rest every few hours and also had to be given the correct amount of water – ten gallons a day. Great care had to be taken in rainy conditions as the yokes could rub their necks. To avoid the heat of the day they would start early in the morning, rest the cattle during the hottest hours, when other chores could be done and then another evening shift would be undertaken, or even trekking by moonlight. The end result, if conditions were good, was still only 15 – 25 miles in a day, and frequently far less. The discomfort of sitting on un-sprung wagons being dragged across rough and often stony countryside must have been difficult to bear and whenever possible even the women and children walked.

Unlike the Voortrekkers who were looking for a permanent home, the *trekboers* were almost totally nomadic, searching for grazing for their cattle and making an extra living from the parts of wild animals they hunted, ivory, horns and skins. In some parts of southern Africa the incidence of tsetse flies was so rampant that cattle simply could not be kept.

The Great Trek has always been part of the Afrikaaner folklore – a tale of heroism and triumph over disaster. It undoubtedly was that, but it had its downside of bringing disruption and bloodshed into the lives of the indigenous people. Away from the government, slavery continued and as the supplies of real ivory ran out, a trade in child slaves, referred to euphemistically as Black Ivory, was substituted.

What becomes more and more obvious is that none of this trading and trekking could have taken place on the huge scale that it did without the assistance of oxen, horses, mules, donkeys and for the great Asian routes – camels.

ET CETERA

1. Wheels with spokes from 1,600BC have been found in the Carpathian Basin. In Anatolia they still use solid wheels.
2. Travellers in early C20th found Yarkend in Western Mongolia still had a famous horse market managed by Kirghiz traders where you could buy the tough little Badakhshani horses, illegally exported from Afghanistan, together with opium.

---oooO0Oooo---

CHAPTER 7

THE ONLY DEER

The Only Deer Reindeer
Photograph by Aviemore Reindeer Herd

"The moon on the breast of the new-fallen snow,
Gave the lustre of midday to objects below,
When, what to my wondering eyes should appear,
But a miniature sleigh and eight tiny reindeer."
"The Night Before Christmas."

Clement Moore

One of the most extraordinary of the animals that came to our rescue, as far as transport is concerned, was the reindeer (*Rangifer tarandus*). Literally from the crown on its head to the soles of its feet the reindeer is unique. It entered our lives approximately 14,000BC or even earlier, the only deer to be tamed and domesticated and bred for its products. All other members of the deer and buck families are prone to panic and to resist any form of taming or training.

Reindeer territory now lies across those parts of Northern China, Russia's Siberia and Scandinavia above the 50th latitude. Their cousins, the caribou's terrain runs across Alaska, Northern Canada and Greenland. At one time with the Bering Land Bridge joining Alaska and Siberia they were literally circumpolar. Now inhabiting an area constrained by climate, they once roamed over most of Europe, down as far as the Mediterranean. So important were they that before the system of naming epochs after types of industry was invented, a large portion of pre-history is known as "The Age of the Reindeer". During the Magdalenian period (17,000 – 11,000 years ago) the main animals for hunting in Europe were reindeer, horses and bison. Reindeer provided the most essential food for Man. But the sudden warming up of southern Europe about 12,000 years ago and the snow and ice began to retreat, so did the reindeer.

Many of the other animals of the time, such as mastodons and sloth bears became extinct or moved away to southerly regions. However the reindeer/caribou survived as did the Musk Ox, the Bison and the Elk but they all became much smaller than they had been during the glacial period. Reindeer and caribou are virtually the same animal, but when the Bering Land Bridge disappeared under rising oceans after the last Ice Age, they developed in their separate ways, and now have developed their differences although they can still interbreed. The most Eastern part of Russia is the Chokotka Peninsula, containing reindeer and the most Western part of Alaska is the Seward Peninsula with its caribou – facing each other across the Bering Sea. Caribou are slightly larger than Reindeer and have never been domesticated, only hunted, by the Inuit.

In the Stone Age Man hunted them for meat and the raw materials like hide, bones and antlers. The big males can reach as tall as 1.5 metres and weigh as much as 140kg. Fossils from the early Pleistocene period over 400,000 years ago have been found in Sussenborn, Germany. They survived where other ice age mammals like mammoths and cave bears went extinct. Indeed they were prolific even right down as far as Portugal and Reindeer bones from as far back as 65,000 years ago have been found in great profusion all over Europe. Some of the bones look as if they have been chewed on and some carved, perhaps to make some sort of an early counting device, or maybe for ornament. Drawings of them on cave walls such as Chauvet in France (32,000 years old) are among the oldest works of art ever discovered. In the British Museum can be seen a pair of swimming reindeer exquisitely carved on the end of a mammoth's tusk, dating from at least 13,000 years ago. Archaeological finds in Finnish peat bogs, always an excellent preservative, showed reindeer harnessed to primitive sleds from 5,000BC. More than 1,500 rock drawings have also been found near Lake Onega and on the Kola Peninsula in Russia. And that link is 8,000 years old.

Both hunting and herding are still taking place, but just when man originally began the herding of reindeer is more difficult to ascertain, possibly about 14,000BC. The annual urge to migrate to and from the coast is so strong that it seems, at first, man was forced to join the herd and migrate with them. Their use to the people of northern Eurasia such as the Sami (Lapps), the Nenets and the Chukchi was and still is like a combination of sheep, cattle and horses, all combined in one animal.

A very early description of reindeer comes from the C1st, and is in Julius Caesar's book "Commentarii de Bello Gallico." The rather inaccurate description runs: "There is an ox shaped like a stag. In the middle of its forehead a single horn grows between its ears, taller and straighter than the animal horns with which we are familiar. At the top this horn spreads out like the palm of a hand or the branches of a tree. The females are of the same form as the males, and their horns are the same shape and size."

Certainly, Caesar was correct in that Reindeer are unique in so many ways. Their antlers, their feet, their coats, their eating habits all are particular to them. Many species of other deer are found all over Eurasia and North America and even in the Atlas Mountains of North Africa – in fact everywhere with the exceptions of Australia, Africa south of the Sahara and the Antarctic, but none have the docile tameable nature of reindeer.

The Latin name for reindeer means 'single file' which, when numbers are sufficiently low to allow it, is their preferred mode of moving on or migrating. Now for their own very special features – the Antlers of the Reindeer consist of two prongs – a long pair sweeping back and upwards before branching, and a much shorter pair jutting its branches forward over the face. But the unique thing is that the females grow them every year as well as the males and this is a real mystery as the growth of antlers is usually attributed to male hormones. There is a possibility that they grow them to protect their calves, but reindeer move around in huge herds which must be a deterrent to their main predators – packs of wolves. Another theory is that it might be to compete for food in the winter. But growing antlers is a terrific drain on the body's resources and energy. They grow at an amazing rate when you consider they are a form of bone and only take three to four months every year to become many branched weapons of splendour. But being a form of bone they mainly consist of calcium, which must be very difficult to generate on a vegetarian diet and must deplete the carefully built up summer store of energy. They are given to chewing the discarded antlers which would help retrieve some of the calcium. The male antlers are discarded in the autumn and the female ones in the spring. Today these surplus antlers are gathered every year by the herders and sold to Russians who arrive in helicopters to export them to China and other parts of Asia who grind them up and use them for making medicine and yet another unlikely form of aphrodisiac. Another even more profitable product is from the velvet stage of the antlers which is not only used for traditional Chinese medicine but now in England and the USA. It is advertised as a whole host of remedies – anti-aging, for improving the performance of athletes, as an anti-oxidant and to hasten the healing of cartilage. In New Zealand they also harvest velvet antlers using local anaesthetic but these are from the antlers of elk and moose; they export millions of dollars' worth every year. Buck's horns, on the other hand are made of keratin, the same as horse's hooves, our nails, claws and rhino horns and unlike deer antlers grow only once and are a male only weapon.

Another special feature of the Reindeer is their feet. Reindeer have broad flexible hooves; a long splayed pair of 'toes' in the front and a short pair of digits or dew claws, unusually, reaching the ground at the back. From the side they look not unlike high heeled shoes. Extraordinarily these hooves change with the season. In the summer the feet are soft and spongy and the reindeer can cross soft soggy ground, but when winter arrives the pads under the feet shrink and expose the edges of the hooves which can then gain more purchase in slippery icy conditions. Another speciality is a kind of anti-freeze made of lubricant oils. These winter hooves also enable them to scratch down into the snow for food, especially their favourite "reindeer moss". This activity is known as 'cratering'. When moving and feeding, they make a low continuous barking sound and as they walk they make a clicking sound that can be heard over from over a hundred metres away. This is caused by a tendon slipping across the joint and may be a way of keeping in contact, particularly in bad weather, snow storms, mist and the winter darkness.

Another aid to keeping in contact with each other and avoiding their main predators – wolves – is that apparently they can see in ultra-violet light. Professor Glen Jeffery from University College London has been conducting experiments in the Arctic with special cameras that use u-v light. He found urine shows up dark and this helps reindeer who can read messages from finding urine. A very important food for reindeer is lichen and this with u-v vision also shows up black. Even more unique is that their eyes literally change colour from golden in summer to blue in the mid-winter – another aid to seeing in the darkness of winter.

Now for a description of their very distinctive coats – like all cold weather creatures, such as the Yak and the Husky, these consist of two layers, a soft thick warm under coat topped by a unique top coat of long hollow air-filled hairs, which in turn trap air. In summer their coats are less thick and darker but in winter their coats become creamy coloured and act as such efficient insulation that Reindeer don't melt snow even when they lie on it. Even more extraordinarily their bodies adjust to changes in temperature to keep their body core warm; even if necessary letting their legs get to near freezing point. All together the double coat insulates them so they can survive the icy Arctic with temperatures of -50°C, howling freezing winds and snow that lasts for nine months of the year. Hair grows even over their mouths for warmth. Reindeer are very able swimmers and can manage to cross the wide flowing rivers on their migration route and this outer coat keeps them buoyant in water.

The cycle of the reindeer year is influenced by the seasons. In the spring they have their best feeding with new green vegetation from grasses and sedges to the young leaves of willows and birches. But for the rest of the year they eat mostly ground lichens which are called 'reindeer moss', mosses and seaweed and they can sniff out plants buried metres below in the snow. In the autumn they stock up on mushrooms which are rich in proteins and potassium. The biggest mushroom is the boletus which can weigh as much as 3kg and measure 45cm in diameter. Their noses are covered with thick long hair which acts as a form of insulation when they are feeding. They are rather keen on licking up human urine, maybe because of the salt content. But a strange part of their diet, peculiar to reindeer is that they eat lemmings – those tiny rodents which do not hibernate, also fish and even bird eggs. They have been known to consume dead reindeer and bits of antlers. These supplements to their vegetarian diet may help supply calcium for the growing of their antlers. They happily drink snow.

Compared to the other beasts of burden reindeer may seem a less obvious choice. Indeed they support relatively few people, but these swathes of Arctic dwellers are as dependant on the reindeer for their requirements as completely as the inhabitants of the Himalayas are on the yak. You can easily imagine how it began with Neolithic man and wolf in cooperation controlling smaller herds of them by driving them into a natural cul-de-sac and then later constructing one; indeed stone remains of trapping pits and guiding fences have been found, which could well be remnants of the Stone Age. Then apart from using them for milking and meat and hides they gradually became tameable enough to be trained as draught animals. A Sledge travelling over snow was a much more obvious concept than transport over land, especially before the Wheel.

The Lapps or Sami (as they prefer to be called) learned their trade – the domestication of the reindeer on multiple fronts from observation. The horse nomads of the Altai demonstrated the art of saddling and riding with their horses; the stock raising of Scandinavian tribes gave the Sami the idea of milking; the Samoyeds' sleds pulled by dogs led to reindeer drawn sleighs. So the reindeer became a sort of combination farm, draught and riding animal. Meat was a fourth gift. So it is the reindeer and only the reindeer that will pull our sleighs, carry our burdens and let us ride them, not that their control or the selective breeding by Man is in the same category of domestication as cattle, horses and dogs.

As always about dating, there is great controversy over just when the Lapps or Sami brought reindeer into a form of domestication, but it seems to be in about 14,000BC and possibly in the southern Altai or Golden mountain region of Siberia. From archaeological studies they have deduced that reindeer were then being bred selectively, used for pulling sleighs and also being ridden. The strange thing about reindeer is the combination of docility and independence. Also they are very gregarious, so they may have hung round human habitation rather like early wolves did, looking for scraps. There is a strong possibility that these famous riders, the Scythians, may actually have started with reindeer. Some tombs in frozen Siberia yielded up horse remains buried with reindeer masks, which seems to indicate a connection at least in the minds of the riders.

The reindeer's strong instinct to lead a nomadic life grew from the necessity to avoid overgrazing. At first the Sami hunters, like nomadic parasites, followed them on their annual migration from their winter range in the northern spruce-fir forests to their breeding grounds of the treeless Arctic tundra; a journey of about 2,500 km. Each herd had its own route. Up in the tundra wolves are left behind and grazing is rich. Led by the pregnant females tens of thousands sweep down at great speed to the coast by May. Of all migrating mammals, the reindeer travel the farthest and the fastest. They are known to cover over 3,000 miles in a year and can travel from ten to over thirty miles a day, depending on the terrain, and can keep up an average speed of 8 mph all day, snatching mouthfuls of food on the move. Their top speed is 32 mph.

The vast herds with their elaborate antlers on both sexes look like a moving forest. There is still a great annual migration, with herds of over 100,000 animals journeying north with their pregnant females to get away from the wolves, their main predators, who cannot follow because the ground is too frozen for them to dig their breeding dens. Other predators are golden eagles and wolverines (a very fierce large type of weasel). They also are trying to avoid flies and mosquitoes, as they have a very low tolerance of insects and this drives them away to windy areas. The gestation period is 220 days and so the journey must end before the young are due. The females usually have their first calf when they are about two years old. Newly born young can keep up with the herd at only one day old and even at that age are faster than a human being; they double their weight in 2-3 weeks and by the autumn weigh over 50kg.

Reindeer having to migrate for food means they are not territorial and also very gregarious and this must have made domestication easier. Body resources are conserved in many small ways, only one mating between the sexes and not much wasteful fighting amongst the males. When food is scarce they can actually lower their metabolic rate.

Males try to acquire a harem of up to a dozen females and will lock their complicated antlers with other competing males, which can lead to entanglement and even death. If threatened they can rear up vertically in an "excitation jump" and exude a special scent from glands between their toes as a warning.

Reindeer have a drug habit! There is a magic mushroom with uncertain effects called Fly Agaric (*Amanita muscaria)*, its cap is bright scarlet with white warts and it certainly looks poisonous. The name goes back to mediaeval times and comes from the practice of breaking bits of the mushroom into a saucer of milk to stupefy flies. And there is the clue – it is a strong and unpredictable hallucinogen and intoxicant and apparently has a similar effect as LSD. So keen on this dangerous mushroom were the Reindeer that to round up a wandering herd all the Sami had to do was to scatter bits of Fly Agaric around and they would come running. The Sami also indulged in this semi-poisonous mushroom, having acquired the habit from their Reindeer. The cap would be dried and swallowed without chewing it and then anytime up to two hours later would come first sleep and then visions, followed by great elation and wild activity. However they find it is less poisonous if it has first passed through the Reindeer's digestive system and been metabolized, so they collect the reindeer urine and drink that. The strength of the drug is such that the reindeer will in turn drink human urine for the effect. This seems a more than desperate way to acquire drugs.

The Sami homeland stretches right across the northern parts of Norway, Sweden and Finland, through Russia and right up to the Kola Peninsula and they have lived in those regions for at least 5,000 years. The Kola Peninsula is surrounded by a sea that does not freeze and is therefore full of marine life. And in Norway the husbandry of the reindeer is exclusively in Sami hands, even though only about 10% of the Sami, mainly the mountain ones are now involved. After hunting, came herding and breeding and then the Sami started to train reindeer. They have approximately four hundred words to describe food, tools and other things dealing with reindeer. They used to castrate reindeer with their teeth.

Reindeer vary a lot in height from about 90cm to 150cm at the shoulder but they can carry up to an average of 30kg for 20-30 miles a day. Of course in winter it's a different matter as drawing sleighs makes movement so much easier and reindeer can draw much larger loads for up to 40 miles in a day. Farming in these latitudes is impossible. Now with the population pressures of Swedes and Finns and Norwegians, you can only find Lapps or Sami North of the Arctic Circle. One very strange thing about them, unique in fact, is that they totally lacked the desire for warfare and when confronted by hostility they simply melted away. Rather amusingly a frustrated Genghis Khan wrote that the Sami were the one nation he would never try to fight again. They may have outwitted the Mongol hordes but they are now being forced to join the modern World, mostly to their cost. There is speculation that the Sami hunters may have used tamed reindeer on leashes to help them approach the herds and then they began to run their own herds.

In Scandinavia the Sami now use snowmobiles in winter and quad bikes in the summer to control their herds. And ownership is shown by pieces cut out of the reindeer ears, and even young children 'own' their special ones. But they have to be registered with the EU, which sounds an almost insurmountable regulation.

All reindeer people have basically the same life style – herding, migrating, breeding their reindeer, but sometimes different local habits. For example another tribe is the Samoyeds who already had dogs pulling sleighs and possibly transferred this task to the newly tamed reindeer. They were seen by a C18th traveler hunting reindeer by using captured ones on ropes as cover while approaching the wild ones with bows and arrows.

Two other incredibly primitive Reindeer people are the Tungus who originally ranged across the vastness of S.E. Siberia and N. Mongolia about 2 million square miles. They moved north in the C12th and C13th with their iron axes and bows and arrows and replaced the even more primitive Yukaghirs of Yakutia who actually continued with stone-tipped arrows right into the C17th. They also used two ancient methods of capturing and killing reindeer, the one was to dig pits with stakes at the bottom, and the other was to form barriers to guide the reindeer into a narrow passage. These stabbing sites were owned by different family groups. So Bronze Age people gradually replaced Stone Age ones. The Tungus had one method of improving their stock and that was to leave a few tame hinds on feeding ground during the rutting season as decoys for new males. As a method of capture they also twisted ropes round their antlers to tangle with the wild ones antlers. They still migrate with their herds across the Taiga, which is a mixture of virgin forest and open swamps, and is totally beyond any form of cultivation. The typical Tungu herd consists of two or three buck, twenty to thirty gelded males, thirty breeding does and the same number of calves. No sticks or whips are used on them.

Another tribe – the poor Yukaghirs, in a parallel to the Aztecs, were all but wiped out by smallpox introduced by Russians. Then the Chukchis who were breeding up reindeer herds and killing off many of the wild reindeer fought with them and finally pushed them into extinction. The Chukchi lived on the NE Chukchi Peninsula and were only discovered in the mid- C17[th]. One of their methods of catching reindeer was using human urine as bait. The Chukchi numbering 15,000 have the largest population of the Asian reindeer people. Their name comes from the Russian word Chauchu which was used to distinguish the reindeer people from the sea people the Angallyt.

In Northern Mongolia live the last few families of an ancient nomadic people the Tsaatan. There close relationship with the reindeer goes back thousands of years and may be drawing tragically to its close. They say "If there were no reindeer we would not exist".

The killing of the wild reindeer by the Yakuts, other tribes and the Russians led to the decimation of the wild herds with no thought of conservation, no plan for the future. In just one hundred years the seemingly indestructible vast herds had been decimated. However they have been allowed to breed up again and in the world there are 3 million wild reindeer and 2 million domesticated ones. Reindeer farms now exist in Russia, Finland and even in France and Scotland. Canada has the caribou of course and they are farmed in the USA and even Japan.

In Eurasia different tribes use reindeer in slightly different ways, but basically there is a tradition of mutual dependence. The milk is the richest of any mammal, but apart from milk and transport there are many other products of the animal. Their meat is delicious, tender and has only 4% fat; it is made into meatballs and sold canned in Scandinavian countries; their tendons and sinews together with their hides are used for constructing their cone-shaped tents called Laitck or Lavvu or Yarangas. And of course these wonderfully insulating hides are also made into their clothes, including boots and gloves. Antlers and bones are fashioned into tools; and the fat is used for lighting. In fact they fulfil all the roles that a combination of cattle and horses do further south.

But reindeer alive are still the main form of transport in the tundra. Some of their saddles seem to have been derived from the Mongolian horse saddles and some from the Altai regions and have not changed much for hundreds of years. The sledges too seem to have first been used further south with cattle and horses. Now, in Russia reindeer are trained to draw troikas, the name for sledges drawn by three reindeer. But a more modern vehicle has intruded on the age old herding – the petrol driven 'vezdehodi', whose noise and smell must make a most unpleasant impact. They also scar the land with their tank-like tracks. Other modes of modern transport are intruding into the traditional lives such as snowmobiles, helicopters, quad bikes and scrambler bikes.

After the Russian Revolution, the Soviet collectivisation policy was imposed in Siberia and the Chukchi were organised into 'brigades' and encouraged to farm their reindeer for meat rather than for transport which was taken care of by their Huskies. This may have been in response to the Russian market, but the fact that migration was stopped, plus the lack of protective forests, combined with the terrible winter of 1984 wiped out nearly half the reindeer population. Even in the best of times the wild herds now have trouble existing in territory heavily grazed by farmed herds. Other tribes such as the Nenets have just about had their cultural identity wiped out with forced schooling and the nomad life actively discouraged.

Another problem for the far north has been mining and the development of oil and gas which has destroyed miles of habitat. For example mining for gold west of the Chukotsk Peninsula has devastated such huge areas of the terrain that it is almost impossible for the reindeer to cross on their migrations. One option could be to leave corridors for linking up grazing. Recently concern has been growing about the herd sizes and the use of vezdehodis, these are based on a tank invented in 1915 and their tracks can do long term damage to the tundra especially in the summer. The name means 'He who goes anywhere'. The huge herds described by the Russians in the C17th have declined with the more devastating use of firearms. In just one hundred years the great resources of NE Siberia have all but disappeared.

Another horror is that toxins of the world find their way to the polar ice and inevitably to the reindeer and the local people.

The North American variety – the Caribou, which comes through French from an Amerindian Mi'kmaq word 'Qualipu' which means 'shovel' and, of course refers to those snowshoe-like feet, have a different life. They are bigger than the Eurasian form and carry heavier antlers. The Inuits, (whose former name was Eskimo), used to drive the caribou into narrow clefts to kill or capture them.

This relationship between caribou and the Inuit has remained at the hunting stage and never developed into the dependant close relationship like the peoples of northern Eurasia. In the North Yukon remains of ancient corrals have been found dating from at least 30,000 years ago. These start about five kilometres wide and narrow down into a funnel for the final capture. Another method observed by the American naturalist Edward Nelson (1855 – 1934) was to twist nooses around the bushes where the caribou were known to feed and these became caught up in their antlers and effectively ensnared them. But as the Inuit never domesticated them or tried to conserve them, their relentless hunting plus the arrival of Europeans led to depleted food supplies in Alaska, especially on the Seward Peninsula. Luckily a Presbyterian missionary, the Rev Sheldon Jackson (1834-1909) noticed this was happening. His friend and ally Captain Healy of the Revenue Cutter Service had visited Siberia and observed that the Chukchi people of Siberia had domesticated their reindeer and were using them for travel as well as food. So the two men imported over 1,300 reindeer for the Inuits to emulate the Chukchi's less wasteful use of Reindeer/Caribou. However the Inuits preferred hunting to herding and domestication has never spread. They hunt more caribou than any other big game species, but it is poorly managed. About 17,000 are herded on the Seward Peninsula alone, but over the years due to poor management the size of the herds has fluctuated wildly from 240,000 in the seventies, down to 75,000 then up again. A more modern use of caribou is the regular mail run in Alaska using a sleigh drawn by reindeer.

Researchers, Vors and Mark Boyce at the University of Alberta (Edmonton, Alberta) have been monitoring the decline of the species which at first they assumed was only taking place in Canada and Alaska. Logging, and prospecting for oil and gas have disturbed them radically and their numbers are dropping fast. But the Boyces and other researchers were amazed to find that the semi-domesticated herds in Northern Europe were also in decline. Climate change is blamed as the warmer summers have bad consequences for the cold adapted reindeer. The big worry is that the habitat is changing too fast for the reindeer to adapt.

Reindeer have been introduced into several new habitats. One was into Iceland in the C18th. The first introductions did not survive owing to harsh winters; but the third herd landed in 1850 is happily breeding up and needs the hunting to keep the population down, which at one stage had reached 7,000.

Another venture was to the other end of the world to the Antarctic island of South Georgia off Antarctica about 80 years ago, by Norwegian whalers from 1911 – 1925 to give them a mixture of change of diet and sport. Just three males and seven females were brought in and one hundred years later they number 3,000. It took over two years for the reindeer to adjust to the change of season and the new diet. There was not much lichen so they had to graze on tussock grass, then the two herds, separated by glaciers began to flourish. They were originally intended for the whalers to enjoy hunting and eating the meat, but whaling gradually fell and finally ceased in 1960's. Their increase to over 2,000 has caused the most appalling environmental damage and the birdlife including the penguins was threatened. In 2013 and 2014 the herds were culled except for some 59 of them which were saved by being taken to the Falkland Islands. The aim is to farm them.

Another venture began in 1952 when reindeer were taken to the Cairngorms near Aviemore in Scotland by a Swede called Mikel Utzi, and his wife Dr Ethel Lindgren. Mr Utzi would dress up in traditional Sami clothes for visitors. The Royal Zoological Society of Scotland brought over a consignment from Scandinavia consisting of two bulls, one ox, two young females and three female calves. The ox, Sarek was the herd leader complete with a bell round his neck. They are used as pack animals for hill walks and a special insect repellent for their use was introduced in 1955, which has been a great success as they can be ridden by small children. Then more consignments came in and the herds were moved to higher grounds where there were fewer mosquitos and flies, those deadly enemies of the reindeer. In 1990 more reindeer were moved to Glenlivet estate in Tomintoul as the grazing was better, they browse far less wastefully than red deer and there are now over 150. They are used for street processions and Christmas celebrations and their meat is harvested. In charge of the reindeer now are the Smiths, Tilly has written several books about her reindeer of which 'The Real Rudolph' tells the story of the reindeer in a readable way.

There is even a small herd introduced into Cornwall, at the Trevarno estate; the first reindeer there for 800 years. There are two Santa's Grottoes, one in Kent and one in Cheshire.

The United States have also gone in for them in quite a big way and there are now about 100,000. They have been taken to the Antarctic for the second time by the French who are breeding them on Kerguelen Island and they now number 4,000, and mix in with seals and penguins, but like the other introduced mammals sheep and rabbits they are not doing the local flora any good.

ET CETERA

1. Father Christmas's sleigh has been pulled by reindeer ever since they featured in a poem written in 1823 'A Visit from St Nicholas'. But as males have shed their antlers in the autumn and Christmas is celebrated in December – mid-winter, the sleigh is drawn by castrated males who keep their antlers much longer.
2. Terrible story told of the Bakhynai tribe whose domesticated reindeer were swept away by a vast herd of wild reindeer coming through. The result was starvation.
3. Sautéed reindeer is a favourite dish in Lapland. Reindeer sausage is widely eaten in Alaska. A company in Taunton, Somerset makes paper from the cellulose fibres found in reindeer dung and is called 'Reindeer Poo Paper'.
4. Reindeer meat can be hung on poles and sun-dried for winter use.
5. In the Finnish saga poem the "Kalevala", Reindeer tallow is used on the snowshoes of Lemminkainen to make them snow proof and fleet. But Hisi makes a reindeer to pursue him out of willow

branches, reeds and rushes. The eyes are daisies and the ears are water flowers.

6. In America 'Deer Antler Spray' is advertised as IGF-1 Plus Formula as an anti-aging and aid for athletes and body builders to reach their peak performance. It claims to hold the secret of "prolonged stamina, strength and youth". It is also harvested from antler velvet in New Zealand from elk.

7. The Nenets numbering over 40,000 are the real cowboys of Siberia. In spite of losing land and being pressurized to settle, they still move reindeer each year from winter pasture to summer grazing 620 miles

---oo0O0oo---

CHAPTER 8

THE BIGGEST BEAST

Indian Elephant with Mahout   African Elephants

"The question is…are we happy to suppose that
our grandchildren may never be able to see an
elephant except in a picture book?"
Sir David Attenborough

Man has always had to make do with whatever animal was at hand, and in India, apart from zebu; it was the elephant, our sixth animal. They were a bold choice when you consider their strength, size and uncertain temper; but at least they were tiger proof. Never really bred in captivity they have been neither truly domesticated nor selectively bred but when you consider the freaks man has made of the domestic dog you wonder just what elephantine travesties we might have engineered – pink elephants would have been the least of them. So, although employed by man, elephants have actually remained wild animals, each one individually tamed and for some unknown reason happy to cooperate with us. The main reason we do not breed from captive ones is an economic one; they

148

take about sixteen years to reach maturity and consume vast quantities of food along the way.

Elephants are related fairly plausibly to whales and dolphins, manatees and dugongs. Even tapirs and rhinos seem likely enough but more astonishingly and further back, they are also connected to the horse and the little rock hyrax or *dassie*. At one stage, members of the Proboscides were just about everywhere except the Antarctic, Australia and a few islands.

Considered dispassionately, elephants are curious looking but when you consider some of their forbears such as the *Gomphotheres* which had two huge parallel jutting tusks on the lower jaw and two downward curving tusks on the upper plus quite a short trunk, their design seems more than reasonable. Like so many of our special six they too originated in North America during the Miocene and Pliocene eras from 12 to 1.6 million years ago and some stayed on in South America right until 9,100BC. Then about 5 million years ago they began to be replaced by more modern looking elephants – the mastodons, a version of an elephant which inhabited the Americas until about 12,000 years ago. They are quite different to elephants being much shorter and stockier. Their name, given to them by a French anatomist George Cuvier, refers to their nipple-like teeth, which are for browsing leaves and twigs as opposed to their descendants the elephants which are grazers. They were sent into extinction by the fatal combination of climate change and the Clovis period hunters about 12,000 years ago.

The other recent proboscides were the woolly mammoths. They originated in Africa and entered Europe, then went on across Asia to China about 3 million years ago. Their name comes from a Russian dialect and means rather strangely 'earth horn'. They stood about 4 metres at the shoulder and apart from their hairiness their main feature were their long very curling tusks. Remains of their teeth have been found all over the world showing them evolving into extremely efficient grinding plates of enamel and cimentum which were needed for consuming grass which having silica in its leaves wears down the teeth.

The last hairy species became extinct at the end of the last Ice Age, with the exception of a few surviving on Wrangel Island in the Siberian Arctic until 4,000 years ago. In 1974 a collection of over one hundred mammoth remains were unearthed by accident at Hot Springs, South Dakota. Like the fate of so many other animals, Man who relentlessly hunted them, had pushed them into extinction. The Japanese, the Russians and the Americans, in a team effort, are attempting to breed them back from extinction, but it could be impossible.

Remains have also been found of tiny one metre high pygmy mammoths on the islands of Sicily, Malta and Cyprus, Crete and Sardinia. These remains date from the Pleistocene epoch (25,000 – 30,000 years ago). The finding of these tiny elephant skeletons is supposed to have inspired Jonathan Swift's tales of Lilliput. The finding of the skulls by the Greeks may have given rise to the Cyclops legend – the giant with a single eye. There is a giant hole in the middle where the trunk was. What a tragedy the tiny mammoths and the pygmy elephants did not survive into our times. The latter would have made the most perfect pets, intelligent, trainable and as long-lived as us and small enough to lie on our beds.

The very first true elephants date back 6 million years to the Pliocene Period. Out of about 352 Proboscidae only two have survived so far. The two survivors are of course the African (*Loxodonta Africana*) and the Asian elephants (*Elephas maximus*). Although they have a great deal of features in common they do have some differences. The Asian elephant is actually more closely related to the mammoth than to the African elephant as has been shown by DNA.

The trunk and the tusks are what make the elephant unique, apart from its size. Indeed Shakespeare might have placed them in his 'fantastical' category. Horror stories about elephants circulated in ancient Rome as they were first judged as monsters with a huge snake curled up on their heads. Made from the fusion of nose and upper lip, the trunk certainly is an amazing organ. With its 100,000 muscles it is an extraordinarily versatile tool which can wrench down great trees and pick up enormous weights and be used as a weapon, but it can also caress its baby with the utmost gentleness, or pick up a coin or even a needle for its trainer. But its primary uses are feeding, drinking, and spraying itself with water, mud and dust. The trunk can hold over eight litres and can filter clean water out of muddy. As though that were not enough, the trunk is also used for the senses of smelling and sound production. Their sense of smell is remarkable and coupled with a prodigious memory means they can avoid danger. The

African elephant has two opposing 'lips' at the end of their trunks and the Asian have only one. A study undertaken by the University of St Andrews (Dr Lucy Bates and Professor Richard Byrne) found that the elephants of East Africa remembered not only the colour of their traditional enemy the Masai, but also the smell. Lately they have found the elephant's prowess at smelling to be a new weapon in the armoury of tracking down poachers. The prey going through water puts off dogs following a scent but not the elephant, which will pick it up again from bushes or grass on the other side.

Communication by sound ranges from trumpeting to squeals and squeaks and even snores that may be echoes of tummy rumbling. They also converse with deep noises; this infrasound can carry for miles and is totally inaudible to humans. In the wild male and female elephants live most of their lives apart, but when the female comes into oestrous for those few special days every couple of years, a sort of dating agency is carried on through these deep low-frequency calls.

The ears too convey the elephant's moods – held out wide to make the elephant appear even larger, they indicate aggression; the elephant will then make a noisy mock charge as a warning, but when he really is attacking, he will charge silently, ears back. But the ears also help in controlling the temperature of the body and elephants will flap them to cool down the blood vessels running through them.

The tusks are formed from the upper incisors, and in cross section show a pattern of lines criss-crossing to form diamond shaped areas called 'engine turning' unique to Elephant ivory. But for the elephant they are tools and weapons combined. They are used to dig for roots and salts and even water. They like to drink 70-90 litres a day. The tusks also form part of the working equipment as levers when moving felled trees. Elephants also employ them to rest their long trunks on. And of course they make most formidable weapons. Asian elephants' tusks are much smaller than the African and the females often have none.

Another very special distinction is that in spite of their diet, which is unbelievably inefficient and wasteful, only 40% being properly digested, elephants are one of the most intelligent of animals, bracketed with primates and cetaceans. Most newly born mammals' brains are 90% of the size of an adult so not very much further development takes place. The African elephant brain weighs 5kg and is similar in construction to the human brain. Humans are born with 26% of their brain potential and elephants with 35%; this leaves plenty of room for development. It has been shown that elephants are self-aware, and also suffer grief when one of them dies. Their social life is important, as they are extremely family minded, life being centred round a group of about ten females plus their calves, led by a matriarch. Even after a short separation they exchange great welcomes with trunk touching and greeting noises. They show altruism by helping out with each other's calves and usually one special female, probably a relative, acts as a kind of nanny to a newly born calf. These nursery helps are called 'allomothers'. Calves are born after a gestation of about 22 months and suckle for 3-5 years, gradually moving onto grass and leaves, so it is a long childhood but they have so much to learn. Male elephants come into a sexual condition known as 'musth' roughly once a year and this can produce aggression especially if two are in musth at the same time and driven by male competiveness.

Elephants can move across the bush with amazing stealth, their feet are fitted with large spongy pads that act as shock absorbers. They are actually walking on their toes but the rest of the foot is padded with fatty fibrous tissue. From prehistoric times, certainly in Africa, they have been making paths that were first followed by prehistoric man; but then eventually we came to follow these tracks with our roads. Elephants don't much care for going uphill, although they can climb mountains if necessary. They read the terrain as well as any topographer with his theodolite. The gradient that was used by the elephant and the mammoth would eventually suit the horse and cart and what suited the horse and cart would ultimately be used by the motorcar.

Their ways are so traditional that they will follow them whatever barriers man puts in their way. For example in Zimbabwe, Lake Kariba was flooded right across an ancient route used by elephants. They now still cross the lake with their trunks acting like snorkels and quite undeterred by the water. They love water and after a good bath they usually finish off with a dust spray; the combination mud dries as a form of protection against sunburn, insect bites and moisture loss. Asian elephants are known to swim 30 miles in 6 hours and so have managed over the millennia to colonise islands.

The African elephant has two branches – the Bush Elephant (*Loxodonta Africana Africana*) and the Forest one (*Loxodonta Africana cyclotis.*) The much bigger bush one can measure 3-4 metres at the shoulder and can weigh 4 to 7 tons whereas the Asian elephant being much smaller is 2-3.5 metres and weighs 3 to 5 tons. Their skins differ a bit as the African is a uniform medium grey but the Asian (*Elephas maximus*) can have variable pigmentation through pink to almost white giving rise to the 'white elephant' stories; Asian elephants are also hairier especially when babies, indication of their relatives, the mammoths. Rather appropriately, the ear of an African elephant resembles the map of Africa and that of the Asian elephant, the map of India. The Asian ones are mainly grazers and the African are mostly browsers, but all great eaters as an adult male can eat 660lbs of grass and leaves in a day. The African ones can live up to 70 years and the Asian ones up to 80 years.

Apart from elephants living in the bush and forests, another unlikely habitat in South Africa is the semi-desert of Namibia, in the Kunene region which consists of the unwelcoming combinations of sand, gravel and rocky mountains. This branch of the elephant family has adapted by having smaller bodies, longer legs and lives in smaller family groups. They are also less destructive of trees. There is another group in Mali, forced into desert conditions by human expansion.

One of the most extraordinary excavations in the world was and still is being made by elephants using their tusks, at Mt Elgon an extinct volcano in Kenya, near the Uganda border. This consists of an underground network of caves and tunnels, created by elephants looking for essential mineral salts. These huge caves may have taken 100,000 years to excavate and shows how instructions must have been handed down the generations. Kitum Cave is the most visited, it stretches 160 m into the mountain and is a Masai word meaning 'Place of Ceremonies'.

In Asia they are extremely adaptable as they can exist in every terrain from steaming hot jungles to cool upland forests, and even deserts. In Sikkim, elephant tracks have been found in the snow at 10,000ft.

The use of elephants was paramount in India, where they figure in the Hindu pantheon as a rather fat man with an elephant head called Ganesh. Once man had selected elephants as possible beasts of burden the first problem was to capture them on a large enough scale to make this viable.

The Asian elephant in the Indus valley was the one first 'domesticated' by man. As already mentioned, they are not bred selectively like cattle, dogs or horses. They are not really candidates for serial domestication as they breed so slowly. They only conceive once every 3 or 4 years, and have a gestation period of 1½ - 2 years, and then suckle the calf for up to 2 years. The calves can't work until they are about fourteen; so it makes more sense to capture new young elephants, when required and train them. They have another drawback, being vegetarian, they have to eat an enormous amount of food, which takes so much time that they can only work 3/4 hours at a time.

The adult male can be very dangerous when in *musth*, a Sanskrit word meaning intoxicated, full of lust and excitement and it indicates the male is ready for mating. So young wild calves are caught and trained to work with astonishing willingness for their human handlers. But what strange alchemy causes wild elephants time after time to enter into this service for Man and to create a unique relationship again and again? When you see the awe inspiring African elephants coming through the Bush, you wonder at the effrontery of man capturing and taming such a beast. Would we have attempted to do the same with dinosaurs had they survived into our times? It is tempting to imagine the Scythian hordes sweeping into battle mounted on *tyrannosaurus rex*.

How did it all begin? It must have been with a calf very young but old enough to be weaned, trained, and ridden. Methods of capture did not vary much since Pliny's descriptions until recently. In North Africa one technique practiced by the early Nubians in the Sudan was to drive elephants into deep narrow gorges that acted as a trap. They then killed the adults and kept the adolescents for use in war. So not surprisingly, by the C6th AD elephants had become extinct in North Saharan Africa. In India there were several methods of capture; the most brutal and counter-productive was driving a herd along a path in which a pit had been dug. There used to be a special clan of Muslims of Arab extraction called 'Panikkians' who had made capturing elephants their profession. The techniques varied; sometimes they would creep up on a lone one with a noose and snap it round his back leg, or they would conceal a noose on an elephant trail and tighten it when the elephant walked over it. Another method of capturing large herds was the '*kheddah*' or stockade into which elephants were driven. Rudyard Kipling wrote a full account of this in

"Toomai of the Elephants" but it was far crueller and more wasteful than his description. A popular and more humane method is the '*mela shikkah*', a way of enticing wild elephants with food and water and '*koonkies*' (tame elephants) to reassure and control them. Where we in the West use the maxim "Fight fire with Fire", the Tamils have the saying "Use elephants to capture elephants!" To get an elephant born in captivity, sometimes a captive female is left out when in oestrus to be covered by a wild male. Nowadays, of course, shooting with a special dart gun is the method favoured certainly in Africa when elephants have to be relocated.

Elephants figure hugely in history – mainly the Asian ones and for centuries as the ultimate prey for the hunter. Just to give a few examples… elephants were known to the Sumerians, and the Assyrians. Elephants in Syria (cut off from the Indian ones) were hunted by Thothmes III of Egypt in C15th BC as relaxation from war. Then in about 1500BC, an Aryan invasion brought the Vedic Age (the time when the Hindu scriptures were written), the kings and chieftains of the more advanced Aryan civilisations had stopped killing elephants for meat. Well ahead of their time, Elephant sanctuaries were actually set up and a manual of statecraft was brought out and acted upon from 300BC to 300AD, which advised the removal of elephants from settled areas and the sending of them to a preserve in the forest and hill areas, patrolled by guards. Anyone caught killing an elephant was put to death. Eventually this area was conquered by the Mughals in about AD1500, and then taken over by the British Raj, which lasted into the C20th. Elephants were important throughout all those 5,000 years. Although used for war in the Mauryan Age they were superseded by horses.

There was a tremendous claim by Tiglath Pileser I who, in about 1100BC on one of his campaigns to enlarge the Assyrian empire crossed the Euphrates and managed to fit in a spot of game hunting. He boasted he had killed 10 mighty bull elephants and captured four. But as he also claimed he had killed dozens of lions and a sea monster called a 'nakhiru' so it's rather difficult to know how true his stories are. Ashurbanipal II 840BC is also supposed to have killed 30, so it was obviously the grand macho exercise for rulers to perform to prove themselves. Libyan elephants were smaller than Indian ones and were domesticated by the Ptolemies of Egypt and the Numidians for Carthage and Rome, but were probably extinct by 600BC mainly from hunting for the ivory trade.

Elephants are also mentioned in the Gajasastra, a Hindu document in verse from over 2,000 years ago, probably C6[th] – 5[th]BC. The name Gajasastra is Sanskrit for 'elephant lore' and it records details of elephant distribution, which was of course over most of India in those days. It also portrays the life and habits, coloured by a certain amount of exaggeration and imagination, and gives instructions on capture and training. Even then there was great trouble over elephants devastating the crops. The Gajasastra tells of this destruction by elephants in the Kingdom of Anga ruled by King Romapeda.

Other references feature in The Mahabharata, which with its 74,000 verses is longest epic poem in world. The King of Orissa from the C12[th] was known as Gajapati, the King of the Elephants. And Orissa boasts the most fabulous Konark Sun Temple or Black Pagoda, from the C13[th], which is one of the 7 wonders of India and a Unesco World Heritage Site. Sandstone has been carved into a mass of animals and 2,000 extremely lively elephants are carved round the base of the main temple. At the entrance two giant lions are crushing two war elephants to death.

More on the practical side is the claim that, in about 2,350BC, there lived in northern India or Kashmir a man named Shalihotra, who is reckoned to be the father of veterinary sciences. No distinction was made at that time between the treatment of men and animals; vets and doctors were one and the same thing. And because of their importance in warfare, particular attention was paid to horses and elephants and their diseases and wounds. Shalihotra wrote a sort of encyclopaedia called the 'Shalihotra Samhita' in which he described in amazing detail equine and elephant anatomy, medicines recommended including preventative measures and even surgical procedure of every sort.

To return to history – the multi-talented Akbar the Great (1542-1605) the 3rd Mughal Emperor of India who ascended the throne at the age of 14, held elephants in great respect and was renowned for being able to ride any elephant however wild or even if it was in musth.

One of the most unlikely linkups is that of Aristotle (of Stagira 384 –322BC), Alexander the Great of Macedonia and elephants. There are more questions than answers. Aristotle definitely tutored Alexander from the age of thirteen and taught him to have an interest in everything including all animals. Aristotle wrote his 'Historia Animalium' and described elephants in amazingly informative detail. But did he actually see a real live elephant? Did he go to India on campaign with Alexander? Or did he only write his treatise on hearsay from other people? Another rather extraordinary authority on elephants was Pliny the elder. Perhaps not so surprising as, at the time, he was the authority on so much, having written thirty-seven books on just about every subject in his Natural History, from mammals and birds to farming and wine making. In Volume III Book 8 on land animals he describes elephants in some detail, including their anatomy and the methods of capture. He asserts, "Elephants are receptive to love and renown and possess the virtues of honesty, consideration and justice to a higher degree than the majority of men." This seems not unlikely. Another quote is "The elephant is the largest of them all, and in intelligence approaches the nearest to man. It understands the language of its country, it obeys commands, and it remembers all the duties, which it has been taught. It is sensible alike of the pleasures of love and glory, and, to a degree that is rare among men even, possesses notions of honesty, prudence, and equity; it has a religious respect also for the stars, and a veneration for the sun and the moon." He also reported "elephants are scared by the smallest squeal of a pig". This claim of 'swinophobia' was borne out as we see later at the siege of Megara. And perhaps it is also due to him that a fear of rats and mice or 'murophobia' has always been attributed to elephants. Apparently that is not true.

Leonardo da Vinci also wrote a paean of praise to the Elephant, which comes amazingly near the truth. "The great elephant has by nature qualities which rarely occur among men, namely probity, prudence and a sense of justice…it never injures others unless provoked. They have a great dread of the grunting of pigs and they delight in rivers. They hate rats." So he too attributed this fear to elephants.

Hunting is one thing with elephants but training them to carry people, to work and to go to war is another thing all together. Once caught then came the training. Elephants need a mahout and a charkaatjie who is the man with the mammoth task of cutting fodder for them. One trusty method was to put a young boy and a young elephant together in a lifelong partnership. The boy would grow up to become a *mahout* in India, or *oozie* in Burma (Myanmar) or *kao-chang* in Thailand, and the relationship between animal and boy would become one of the closest that can exist between man and beast. The elephant would start its training at about five years old and would learn to obey words of command spoken in a special dialect derived from Sanskrit, which is the ancient language traditionally used for all verbal communication with elephants. They are also given signals with the *mahout's* feet touching the back of the ears. At the word of command an elephant will lift its *mahout* effortlessly onto its back. I watched amazed when an elephant we were riding in Nepal was told to pick up the *ankus* (a 5 – 6 foot long wooden pole with a metal point and a hook used for controlling elephants) and searching around found it in the eight-foot tall grass and handed it back. Sadly the age-old traditions of the *mahout* are dying out and the over 16,000 elephants in captivity are desperately needing help. This is in spite of the Indian Parliament passing a Wildlife Protection Act in 1972. The profession of *mahout* had, unfortunately, never acquired the rather absurd glamour of the American cowboy.

Elephants are used for work in many different ways. They were trained as beasts of burden as they can carry loads of 300-600 Kilos for up to fifteen miles at 4 mph over the most hostile terrain, mountains, rocks, and through deep grass. Helping with building became another use for elephants and one historical example dates from 1399 when the Mongol conqueror Timur the Lame used 95 elephants to help build the Bibi Hanim mosque at his imperial capital, Samarkand in Uzbekhistan. Logging was yet another use of the elephant and this increased on a large scale during colonial times, in about the 1850s and continued into modern times. They are still used for logging especially in Myanmar and can haul twenty-foot sections of wood, weighing up to four tons through the jungle and drop them into the river, to be floated downstream. Even in Sri Lanka there are over 500 elephants being used in work. They are unique as beasts of burden because of those handy tools, the trunk and tusks. Camels, oxen, horses and all the others can merely carry their burdens or act as draught animals but the elephant, with its superior intelligence can do so much more. So they were essential for all forms of transport, carrying every sort of load, even drawing carts and ploughs. They were and still are in some parts of Asia, used in rice paddies, for drawing water carts and to harrow coconut plantations. A rather interesting exercise was made in 1927 – a comparison of costs was drawn up of different ways of ploughing 100 hectares on a cotton station. With a tractor the cost was 704 Belgian francs, with oxen 156 francs, with 50 men 111 francs but with elephants only 102 francs.

Elephants have a prominent backbone ridge, which is extremely vulnerable to heavy pressure so before any kind of saddle can be put in place, the spine has to be protected first by a pad of cloth (*guddela*) and then by a cushioned pad (*guddee*) with a central groove to fit over the ridge. Over this is placed an iron saddle with girth rings and ropes to go round the neck and under the tail to hold the saddle in place. The whole lot is finally topped with a howdah or panniers for carrying loads or more modern bench seats for carrying tourists. Bulldozers and 4 wheel drive vehicles are now gradually replacing much elephant work. But if the price of petrol rises much higher it could become economic to bring them back into production. In fact during World War II, the Nazis, desperately short of petrol, used elephants from the famous Hamburg zoo to plough their fields.

The overall population of elephants in Asia numbers 35,000 to 50,000. There are still 12,000 elephants in Thailand and probably the largest number is in Myanmar, which has the least damaged population of working elephants. It seems the Buddhist oozies have a closer emotional bond with their charges than the Hindu mahouts, who come from one of the lowest castes and suffer from low esteem. In fact India, Thailand and Myanmar together own 80% of working elephants. Sri Lanka has 2,500 to 3,000 elephants, of which about 500 are working elephants. Tourism is on the increase in China and in the tropical rainforest in the far south of Yunnan Province near the Burmese border lies the Wildlife Elephant Valley of Xishuangbanna. There they have a small treetop hotel from which you can watch up to 70 wild elephants and other wildlife. In Nepal at Chitwan National Park elephants are used to carry tourists to view rhinos in the three metre high grass. Also on the tourist front, there was a rather fascinating occurrence during the Tsunami of December 26th 2004 – elephants were carrying tourists along the beaches of Sri Lanka then when the water receded the elephants wisely and firmly took off for the higher ground ignoring all commands from their mahouts. Of course their behaviour was justified when the huge surge returned.

The main drawbacks of the working elephant are that they need so much food and feeding time. An adult bull consumes at least 660lbs a day. For every 3 days on they are given 2 days rest, and in Myanmar they have a working year of only nine months. No trades union could organise more lenient terms. Although Asia was the main home of the tamed working elephant there have been experiments made in Africa. As will be seen later, African forest elephants did figure in warfare. But in more recent times King Leopold of the Belgians organised training of African elephants in the Congo. He started by shipping four Asian elephants in 1879 to be landed on the coast of Tanganyika, to the amazement of the locals. Under the command of the British ex-consul to Baghdad the elephants with 13 mahouts, 8 soldiers and 700 porters were sent into central Africa. It was a complete disaster as the elephants died on the way and a local chief massacred the rest of the party. In 1899 Leopold tried again. Commandant Jules Laplume caught some African elephants from North of Zaire, and founded the "Service de Domestication des Elephants". They tried various methods to capture the elephants and finally settled on shooting the mother then catching the young calf with bare hands and ropes. This method was used for all of 58 years before at last a more civilised way was found to drive off the mother instead of shooting her. Older tame elephants, the equivalent of *koonkies* were used to calm the young captives who were usually around 15 years old. By the year 1913 the centre had 33 elephants. The centre nearly closed several times but was rescued by Leopold and by 1920 he had imported 7 mahouts from Sri Lanka.

What were the other uses Man could find for elephants...for war was the obvious one, but that is dealt with in Chapter 12. Ivory was and is a tragic one and there had always been a demand for it. Just to take a few famous examples, in 1352BC, Tutankhamen was buried with a casket inlaid with 45,000 pieces of ivory; and in 1000BC King Solomon had a great throne built of ivory and gold. But by the C19th Man's greed for ivory had reached the most damaging proportions and as automatic weapons were now freely available to poachers as a result of the endless civil wars of Africa, this combination led to the most terrible decimation of the species. In 1980 there were estimated to be about one million elephants in Africa and in only ten years more than half had been killed for their tusks. Three quarters of ivory had always been used for carving in the Far East; it was also used for making billiard balls and piano keys. Mammoth tusks had also been used but the quality was not nearly as good and many mammoth tusks owing to their great age were actually rotten. At last sense began to prevail with the banning of imports of ivory first by the USA then West Europe and finally and reluctantly by Japan. In October 1989 CITES

(Convention on International Trade in Endangered Species of Wildlife Fauna and Flora) voted for a universal ban. The ban coupled with demand toppled the price of ivory. In Kenya, for example, the price dropped from $50 per Kilo to $3. This led to a dramatic falling off in poaching. Growing tourism also gave the local population good reason to conserve their elephants. Camel bones are now quite a good substitute for ivory.

Botswana, Zimbabwe and Namibia wanted to lift the ban ten years later and sell off on a one-time basis, large piles of ivory, accumulated from elephants that had died from natural causes and confiscated poacher's ivory. This they argued would pay for maintenance of game parks. Arguments against and for a certain amount of culling move to and fro and every country has its own approach to problems of over population. There is fierce debate on the decimation of woodland. Is it part of a long cycle? Unlike cattle, elephants, in the very long run, probably do not cause desertification.

Elephants have always been presented to one potentate from another, certainly making it a present with a difference, the equivalent of a "must have" car or yacht of today. In 797AD the dark civilised Sultan Haroun al-Rashid the Caliph of Baghdad, made diplomatic overtures to the almost illiterate blonde giant, Charlemagne, by giving him spices, perfumes, an elaborate water clock and an elephant called Abu l'Abbas, the first elephant to be seen north of the Alps since Roman times. Charlemagne rode him into battle against the Danes in 804AD. Unfortunately he drowned crossing the Rhine, which is strange as elephants are usually good swimmers. Four hundred years later, in the C13th the Sultan Al-Kamid in Cairo sent an elephant to King Frederick II of Sicily who was known as *Stupor mundi* for his knowledge of nine languages. The elephant lived for decades and caused a sensation in a triumphant procession through Milan, celebrating victory over the Lombard cities. He emphasised the victory by pulling the captured Lombard flag carriage. When he died in 1248 in Cremona, the local people fully expected the elephant bones to turn into ivory.

Another elephant story with a less happy ending came after the failed 6th Crusade in the C13th when King Louis IX of France brought back an elephant with him and then sent it on to his brother-in-law in England perhaps fulfilling the Tamil curse. A special house was built for it at the Tower of London but he didn't survive long.

There is a wonderful drawing of an elephant by Mathew Paris, the famous C13[th] Benedictine monk who lived in St Albans. He wrote a book called *Chronica maiora*. On the elephant's neck sits a man, ringing a bell and on his back is a box, correctly attached to the elephant, carrying about a dozen men blowing trumpets, clashing cymbals and driving the elephant forwards with spikes.

In 1515 an Indian elephant called Hanno was presented to Leo X, the Medici pope by the King of Portugal, Manuel I. He was ridden into Rome by a magnificently clad Moor and forever dispelled the myth that elephants had no joints, by kneeling to the Pope. There are various theories about the cause of Hanno's death – drinking too much red wine, falling into the Tiber or most improbable of all – being coated with gold leaf.

The Brenner Pass between Italy and Innsbruck has several inns named after Suleiman the elephant who, to local astonishment, marched across the Alps through the depths of winter in 1548. This greatest of all designer animals was a present from King John III of Portugal to Maximilian the Crown Prince. He was then transported by boat along the River Inn and the Danube and into Vienna. Apparently in the excited crush of the crowds a child fell in Suleiman's path who picked it up with his trunk and handed it gently to his mother. This all caused waves of elephant mania. He was later used in Maximilian's coronation and after he died, some of his bones ended up made into a chair that still stands in Kremsmünster Museum, an unusual commemoration to a famous elephant.

As I've tried to demonstrate, elephants entered our lives in many roles and not just as beasts of burden. Their trunks and tusks as tools, combined with their intelligence made them extremely useful and they played many other parts in our lives as later chapters demonstrate. They also have shown they can have a close friendship with humans. Lawrence Anthony took on a herd of rogue elephants in Kwa Zulu South Africa when he was running a game park called Thula Thula. They amazingly became his friends and he just as amazingly encouraged them back into the wild. When he died recently the two herds journeyed back to his house which they had not visited for a year or two. Their trek lasted about twelve hours so they were some distance away when he died. They hung around in mourning for a couple of days and then drifted away. How did they know he had died?

ET CETERA

1. The trunk has over 30,000 muscles & holds 12 litres of water and can filter clean water out of muddy. They create vital pathways through forest or savannah and can be the ultimate off road vehicle.
2. A cousin of the King of Siam visiting London and meeting Queen Victoria paid her the highest compliment in his book and told her she looked like a beautiful, majestic white elephant! History doesn't relate how she received this accolade.
3. Elephants, when wanting to raid a banana plantation, have been known to stuff mud into their bells to prevent them ringing and betraying their movements.
4. When a cousin of mine was farming in Kenya in 1950 an ancient elephant track ran through part of his farm. One day he was walking near the track with two new little terriers when along came a large male elephant. Before Robin could do anything to catch them the dogs had flung themselves barking at the elephant who swept his trunk around to sweep them into oblivion but they evaded every swipe and finally the poor elephant, had to admit defeat and retreated.
5. Russian geneticists are trying to breed back mammoths to farm for meat. They reckon one mammoth weighing 4-5 tons would produce enough meat for 100 people for a year. They are working on resurrecting DNA from hairs found of a mammoth found at Yukagir.
6. "The Greek shall come against thee,
    > The conqueror of the east:
    Besides him stalks to battle
    > The huge earth-shaking beast,
    The beast on whom the castle
    > With all its guards doth stand,
    The beast who hath between his eyes
    > The serpent for a hand". Thomas Macaulay (1800-1859)
7. A perfectly preserved month-old mammoth, who had died 42,000 years ago, was found in 2006 by reindeer herders in NE Siberia. She

has been given the name Lyuba and is the most intact mammoth carcass ever found. Even her eyelashes were preserved.

8. A final word from Arthur Schopenhauer, the German philosopher who declared, "The Idea of an Elephant is imperishable!"

---oo0O0oo---

CHAPTER 9

CARRYING OURSELVES

Carrying Ourselves. African Woman with bag on her head.

Mother Carrying her Baby

But Man has never entirely been able to lay all his burdens on the backs of other animals and walk off into the sunset. Even after he domesticated the extraordinary and disparate animals that came his way, there would always be circumstances in which he still had to load up himself. Men and women in Africa particularly, carry extremely heavy weights, not on their backs but balanced on their heads. The porters who were used by the great explorers and missionaries, the hunters and the colonial officers, all used this method. Famous names such as Livingstone,

Burton, Speke, Burchell and Selous went on Safari into the African interior. These journeys were gigantic, lengthy and dangerous undertakings. Their expeditions consisted of gun bearers who had to be trusted, interpreters who were local and temporary, and literally dozens or even hundreds of porters; the whole entourage controlled by a Headman. Horses, mules or donkeys could not be used because of the deadly horse sickness; however these porters were capable of carting 60lbs for 10 – 15 miles a day. One or two tribes, such as the Wanyamwezi one of the largest tribes of Tanzania, actually specialised in providing reliable porters.

Mountains and high altitudes can cause other problems. So it was men who had to be the main beasts of burden, for example, in Chile, a country stretching 2,500 miles from North to South, and with much of it rising to over 12,000 feet. The fuel that kept them going, through cold and high altitudes, through hunger and thirst came from the Coca plant, the leaves of which were used by noblemen and porters but never by women. This rather innocent forerunner of cocaine was chewed as a wad of leaves to enhance performance by preventing drowsiness and helping the lungs to extract more oxygen from the rarefied air. When chewing the leaves, men could carry their own weight, which was double the amount a man could cope with at the high altitudes without using coca. The average man in the Andes was 5ft tall and weighed 100 lbs. So in its unrefined state the coca leaves acted as an essential fuel. The leaves were also chewed by top couriers called 'chasquis' who carried vital messages, decrees etc; these men were born at a high altitude and trained from an early age. These Andean porters used to start the day with porridge then go for twelve hours on their journey chewing constantly and changing the leaves every forty-five minutes. On this fuel they could cover 20-25 miles in a day over steep territory. So wedded to this drug were they that a day's march would be described not in miles or hours but by coca chewing periods called 'coca'. They even fed infusions of the plant to their llamas and later, when the Spaniards had introduced horses, to them too. Now highly refined and marketed as crack cocaine it has become a dangerously addictive drug, but chewed or made into tea it increases pulse rate, appeases hunger and prevents drowsiness. When I visited Machu Pichu it was available at the hotel for those who were suffering from altitude sickness. It had a fabulous effect on a busload of grey haired female American tourists.

Everest, K2 and all the famous peaks in the Himalayas and all over the world, the Alps, the Andes, Mt Kenya and Mt Kilimanjaro were only conquered by the climbers carrying their own baggage up the final pitches.

In the Himalayas it is the Sherpas who carry huge loads for all kinds of expeditions – food, tents and equipment of every kind for mountaineers, tourists, and scientific explorers. Many places simply cannot be reached by helicopters, which anyway are prohibitively expensive. Yaks and Dzos (hybrids of cattle and yaks) are used wherever possible but when it comes to climbing the high rock faces they obviously cannot manage. Curiously, many of the Sherpas are lowland farmers who come up to the high altitudes to earn extra money. They have developed a special strap that fits over their head and supports the weight on the spine. They then manage to balance the load by bracing the strap with their hands. But the amazing thing about these Sherpas who, lacking a strong physique and without the stimulus and staying power bestowed by coca, is that they can carry more than their bodyweight all day at high altitudes. A study by Belgian scientists found that the two main factors that assist the Sherpas are the use of the head-strap, which raises their centre of balance and helps distribute the weight. And the second factor is that for some physical reason the Sherpa metabolism, at high altitudes, increases only half as much as Westerners and that enables them to do far more while expending less energy.

A society to advise mountaineers and explorers how to treat their porters in developing regions of the world such as Tanzania, Nepal, and Peru, was founded in1996 in Boulder, Colorado. It is called the International Mountain Explorers Connection – IMEC. The society gives guidelines to prospective mountaineers on treatment, wages and even clothing for the local porters on Mt Kilimanjaro.

Religion also beckons Buddhist nuns and monks up a limestone cliff above the Terdom nunnery, north of Lhasa in Tibet. The Lama guides the line of porters and nuns up ladders and then crawling through a steep tunnel to a sacred cave at 16,000 feet, there to meditate from the new moon to the full moon.

Having porters is nothing new – the Romans, not only had mules and oxen to work for them they also used the saccarius or male porter, but their use was limited by the amount they could carry. Saccarius means someone laden with sacks.

In the Arctic rim region an innovative way of moving over the snow was developed, vital for hunting especially in winter; and this was skiing. The world's oldest ski 8,000 years old was found in Russia; and skis over 5,000 years old have been found in Norway. Stone Age rock carvings of examples of hunters on primitive skis have been found in peat bogs uncovered by the retreating of the glaciers at Rodoy in Norway. One of these carvings was used as the pictogram for the Winter Olympics in Lillehammer in 1994. Iran was another area supposed to have had skiers using boards made of animal hide dating back to 2,000 BC. Moving into the age of the written word the Eddas, a collection of Old Norse Icelandic poetry, dating from the C13[th] or earlier, celebrated the dashingly fast skiing of King Harald Hadrack (1046 – 1066AD). There was even a ski goddess Skade, whose name was the root of Scandinavia. At first the skis were of different lengths, the shorter one had skin underneath it and was used to push off the longer gliding ski. Also to begin with, only one pole was used. By the C18th the Nordic skiers had developed different skis for different pursuits, for speed, for hunting and for going into the woods, probably for hunting.

Two other far more modern adventurers made full use of skis – the first was Nansen when he and five others became the first men to cross Greenland in 1888. The other was Captain Scott on his ill-fated attempt at getting to the South Pole in 1912. Now, of course skiing is a well known sport and celebrates its own Olympic Games. It provides an enormous tourist industry all over Europe, North America, and even in the Southern hemisphere, in places such as New Zealand, the Andes and even the Snowy Mountains in Australia. Skiing has moved right out of the purely functional means of transport.

Another device for crossing over the snow has been snow-shoes. Even with dogs to pull the sleds, in tribes like the Amerindian Ojibwa the women still had to carry huge loads of over 60 Kilo. The men managed to get off this duty by doing a spot of hunting. The snowshoes varied in design from long narrow almost ski-like ones for beaten tracks, but for fresh deeper snow, round ones were preferred; the frames of these 'shoes' were constructed from birch or willow twigs and the webbing was made from strips of hide.

Now for transport in hot climates – carrying-poles are the preferred method of transport that is used extensively in Thailand and other parts of SE Asia. They are made by cutting a thick bamboo at least 4 inches in diameter and about 58 inches long. It then is split length wise and carved carefully to taper towards both ends, except for a node left at each end. These thicker ends secure the ropes that are attached to carry a pair of panniers. The extraordinary thing is how much weight the slightly built Thais can carry – even the women can manage over 20 kilos. What helps them is that the rhythm of the running motion of the carrier sets off a swinging and bouncing movement of the panniers, and that seems to lessen the load. These flexible poles contrast most favourably with the *pulker* bearers of Bengal whose poles are made of stiff, unyielding wood, and also with the heavy yokes and pails of milk that used to be carried by the bigger and stronger Dutch milkmaids.

Panniers vary in design and usually come in pairs. Their name comes from the French for bread (le pain = bread). In the East special baskets of woven bamboo strips and rattan often have a fitted tray for prepared foods for sale in the markets. And of course as already noted they are used on donkeys and mules and this method of transporting goods has been adapted for use on bicycles and motor bikes in the modern world. Sometimes pairs of panniers of light weight wicker, strapped to a mule were used to remove the lesser wounded from the battlefield especially in the Crimea, but also in the American Civil War. The same devices were also used to carry goods generally and even tourists in the South of France.

Another method of carrying people is by Sedan Chair. This consists of a small upholstered cabin with windows containing a comfortable chair for a single person. The roof can be hinged back to make it easier for the passenger to stand up and climb out. Two long poles are slipped through metal brackets on the sides of the cabin and two porters or more, one in front and one behind; they were mainly used to carry persons of quality through the filthy streets of London. Up to eight men would take it in turns to carry the chair. They were invented as early as 1581 and were used by the Duke of Buckingham, but in an early example of political correctness they were at first condemned as exploiting human beings. However those objections had fallen away by the 1630s when Sir Saunders Duncombe patented his version of the 'hackney chairs' as they were then called.

According to John Evelyn's Diary Duncombe had imported the idea from Naples where they were called 'seggioli'. After that, they became more and more popular so that by the late C18th there were literally thousands for hire all over England, Ireland and the continent of Europe. Apart from being cheaper than hackney carriages they had the added advantage of being able to pick up the passengers from inside their house and deposit them again right inside their destination house, without them having to set foot in the grimy streets and mess up their shoes or their expensive dresses. The chairmen carried the poles on their shoulders and generally two would be sufficient to carry one person at great speed through the narrow streets of London, but in the latter years of the increasingly corpulent Henry VIII, four strong men were necessary to bear his particular weight. There is a lovely neo-classical example of a Sedan chair, made for Queen Charlotte, to be seen in Buckingham Palace.

Some of the well to do owned their own chairs but for the rest, there were chairs for hire on stands like hackney carriages, or today's taxis. And just like hackney carriages they were licensed and had to display a number. The 'chairmen', as they were called, wore a distinctive uniform with a large cocked hat, and often did double duty as unofficial policemen in a public brawl.

In Scotland the Sedan chair was regulated by a fare system from the mid C18th and they were sometimes used as an early form of ambulance. And in Bath the chairs were employed to move invalids or people 'taking the waters' from the mineral baths to their lodgings without them being exposed to the dangerous cold air.

Sedan chairs were used widely in France and there is a division of opinion over whether the chair was named after the French town of Sedan…but most think not. They were also in use in America especially in Boston and New York and are known to have been employed by Benjamin Franklin.

Another more elaborate version of the sedan chair was the 'Sedia gestatoria', which was really a luxurious portable throne for the Pope, complete with silk upholstery and even gold rings for the carrying poles. It was borne on the shoulders of twelve men in red uniforms and accompanied by two huge ostrich fans one on either side. The tradition is thought to date back to Byzantine times or even to Roman times when leaders and newly elected consuls were carried on a 'Sella curulis' through the streets. Of course nowadays the Pope rides in a Popemobile.

In India the equivalent of the Sedan chair is called a 'Palanquin' or 'Palkhi', and as it could be curtained, one of its main uses was to carry women, particularly Muslim women in purdah. In Korea the Sedan is called a 'Gama' and is much more use than wheeled vehicles owing to the lack of paved roads. They were also very much used in China, and still are for tourists in inaccessible places. But the most elaborate form in China for people of great importance consisted of virtually a small room carried aloft by at least a dozen bearers. There's a marvellous picture of the Qianlong Emperor Xu Yang on his Southern Inspection Tour 1764 – 1770 in a litter carried by no less than 16 bearers. Then during the Han Dynasty (206BC – 220AD) people were carried, rather less elaborately, in chairs strapped back to back on the carrier, like a rucksack. In Japan these were widely used by the elite samurai as there was a shortage of horses. It must have looked rather strange an adult being carried like a baby.

A sort of early version of the sedan chair was the litter, which was used in early Egypt to transport the rulers and also sometimes the state idols. In Ancient Rome the elite were borne, usually by four slaves, reclining in litters called 'lectica'. It looked very much like a small four-poster bed and must have been extremely heavy to carry even without the passenger.

Another problem has always been transporting the wounded from a battlefield, because if badly done, death or worsening of the condition can result. And so stretchers were contrived in various forms, the most usual being a length of canvas slung between two poles. These were also used for carrying ill or injured people from ambulances into hospital. An interesting version of this is the Neil Robertson stretcher, which was originally designed for the Navy. It was formerly made of bamboos, joined by canvas. The wounded man was rolled up in it then held in with cotton straps with buckles, making a tube. Finally a rope attached to the head end of the stretcher allowed it to be pulled vertically up through a hatch or other narrow space. It is useful dealing with such diverse situations as out of submarines or in mountain rescue, into helicopters or when potholing. The modern version has an aluminum folding frame and can be carried in a compact shoulder bag. On the battlefield and in the absence of a stretcher there was an impromptu version made from a blanket doubled up with slits at the corners big enough for a musket barrel. One musket was then threaded through the fold in the blanket and the other through the end slits to produce a rough and ready litter.

So now there were varying devices for carrying kings, the rich and the Pope. Also ways of moving the sick, injured and the dead were organized. But who were the human beings that had to be carted around more than any of any of these? Babies...thousands and then millions of them had to be carried by their mothers, by their nannies and even sometimes by their fathers. Babies have been moved about in many ways, the oldest and most obvious being the shawl method, in which the shawl is knotted in front and the baby is carried on its mother's back, with breasts holding up the shawls' knot and the backside of the mother helping support the baby; that has always been the African way. The Chinese also use a cloth to carry their infants but it is tied more elaborately over the shoulders round the child and then round the mother's waist. They carry them on their backs or on their fronts.

Amerindians used the papoose, which was actually an Algonquian word meaning child, but which came to mean the child carrier. The papoose is the antecedent of all the modern child carriers that you see slung sometimes on the back and sometimes on the front of the parent.

Then the wheel was invented, an event which totally transformed mobility. One of the earliest combinations of wheel and man power was the Chinese Wheelbarrow, which seems to have been invented during the Han Dynasty (206 – 220AD) and the earliest surviving picture of one is on a tomb-shrine in Szechuan province. It evolved into several versions. The obvious one consisted of a large single wheel in front of a platform to bear the load and handles projecting to the rear. The operator would pick up the heavy end and push, thereby combining the lever with the wheel. Another type was invented by General Chuko Liang (181-234AD) which had its large single wheel encased in housing with a flat platform on both sides. This carried loads or wounded soldiers at a height similar to a rickshaw and was kept a closely guarded military secret. It had the added advantage of releasing pack animals for other purposes. It was built for carrying heavy loads a long distance, so it sometimes had a second operator in the front. The two main versions became picturesquely known as "The Gliding Horse", operated from the rear and therefore pushed. The other was known as "The Wooden Ox" and pulled by handles from the front. The Chinese even tried a 'land junk' with a sail to waft it along. In 1974/5 a priest from Manchester, called Geoffrey Howard actually tried to sail a prototype of this wheelbarrow/junk across the Sahara, with a certain amount of success.

The wheelbarrow appeared in Europe during the Middle Ages, but whether the idea was imported from the East or invented independently is not known. There is an extraordinary depiction of one on a stained glass window in Chartres Cathedral, of all places, dated from the 1220's making Marco Polo (1254-1324), the Venetian traveller's claim to its introduction, like so many of his claims, a bit dubious.

The European version of the wheelbarrow consists of a bucket or bowl, stands, two handles and of course the wheel, and is mainly used for gardening and construction work. Lately the wheel has been transformed into a large ball, invented by James Dyson. Apart from the obvious advantage of greater manoeuvrability, it doesn't score wheel marks into lawns, or get sunk in sand.

Another wheel plus man propelled invention was the Rickshaw or Jinrikisha. Legend has it that it was introduced in 1869 by an American missionary, Jonathen Scobie, who used it to take his invalid wife about in Yokohama. Another theory is that it was invented in Tokyo by three Japanese men who started building rickshaws. Its name arose from a combination of Japanese words *jinrikisha* jin = man, *riki* = power, *sha* = vehicle. Not surprisingly they rapidly caught on as they were faster than a palanquin and using man power was cheaper than horse power. By 1870 there were 40,000 odd in Tokyo alone. They soon spread into other parts of Asia. Bangladesh was even nicknamed "The City of Rickshaws". By 1880 they had reached India, now they have been largely supplanted by Pedi cabs (or velotaxis or trishaws, as they are variously named.) as they were banned in 2006 and branded as inhumane. In Shanghai and China generally they were banned by Mao in 1949 as 'symbols of bourgeois imperialism'. Another country where rickshaws caught on was South Africa; they were introduced by a British sugar magnate towards the end of the C19[th] and by 1904 Durban had over 2,000 rickshaws pulled by Zulu Rickshaw Men who wore gigantic headdresses, decorated with horns, beads and plumes.

Wheels in some parts of the world also came into use for the endless task of transporting babies and children about. Perambulators or prams were invented and in London, Hyde Park and Kensington Gardens were alive with nannies wheeling their charges and showing off, not only the child but the latest model. Silver Cross, from a firm dating back to 1887, was considered the most elegant – the Rolls Royce of the nursing world. Now there are all kinds of carry cots and push chairs, strollers and the American word 'buggy' has become ubiquitous. Finally for transporting children in cars there have been special car seats invented, that in the United Kingdom are literally tied up in health and safety laws.

Another very useful method of carrying goods, gear for hiking, climbing, camping is the Rucksack. The word comes from the German meaning 'back' but is used in the West. It is a kind of haversack but instead of being carried by a single strap it is mounted on a lightweight frame to distribute the weight better. Now often referred to as a backpack especially when referring to the new cult of 'backpacking' which usually means far flung travel by young students in their 'gap year' before they settle down to university. In 1902 the Rucksack Club was founded in Manchester and had over 400 members. It was formed to encourage walking tours.

An extraordinary thing concerning the transforming wheel is that it had actually also been invented in Mexico but was found only on ceramic toys and not on any type of vehicle. True, they had no domesticated animals but you would have thought that conveyances like rickshaws or wheelbarrows would have been extremely useful, but somehow they never made the connection.

With petrol prices and congestion getting worse, the Pedi cab, an offshoot of the rickshaw is staging a comeback in London, New York and other cities as a form of 'green' approach as well as being a crafty way to cheat the traffic jams by various dodges such as jumping pavements.

So down the millennia we went staggering on, bearing some of our own loads, on backs, on heads, in chairs or once we had invented wheels, in prams, wheelbarrows and rickshaws but there is a limit as to how much we can manage in weight and in distance. Luckily four-legged help was at hand.

ET CETERA

1. One famous adventurer who went on a most expensive safari was Theodore Roosevelt with his son Kermit. It cost £15,000, which

in 1909 was a great deal of money but he did gather hundreds of specimens for the Smithsonian.

2. Two African safari porters, Chuma and Susi deserve special mention, for love of their Bwana they carried the dying David Livingstone firstly in a *kitanda* or stretcher through pitiless rain. This was 1873 in the south end of Tanganyika. Then after he died they wrapped his body in sail-cloth and bark, like a mummy and lashed it to a pole. For nine months they carried it often through hostile territory to Bagamoyo on the coast. There the British Consul arranged for it to be shipped to England and its final resting place in Westminster Abbey.

3. When Scott of the Antarctic's horses died on his journey back from the South Pole, the men had to carry the baggage and as they could only carry about half the weight, they had to do three times the distance to keep their supplies with them. Special horse shoes for snow had been available but Scott had been dissuaded from using them by Captain Oates.

4. In 1200AD Tseng Min Hsung a Chinese man, apparently and colourfully boasted about the wheelbarrow that could cope with "ways, which are as winding as the bowels of a sheep will not defeat it."

5. A retired Royal Marine told me that on D-Day a fellow RM officer was so badly wounded in the leg he was unable to walk. As hostilities were still raging it would not have been easy to carry him but luckily they found a wheelbarrow in someone's garden which they used to push him along in until they found a dressing station.

6. In the Zermatt museum you can see a sort of wicker rucksack made by the women in winter for carrying hay, wood, green fodder, dung and maybe even babies. In the very local dialect they are known as Tschifera.

7. I tried carrying my son like the African women do, on my back with a shawl tied round my chest to support him but found the European embonpoint was not suitable for supporting either the shawl or the baby and he ended up in the King's Road, Chelsea halfway down my calves!

---ooo0Ooo---

Chapter 10

MYTHS, MAGIC & MORES

Myths, Magic and Mores. Ganesh, the elephant god of India.

Man's attitudes to animals are incredibly complicated; they vary from country to country, from century to century and from person to person. One man's god is another man's dinner. Very early on in his development man began looking for something beyond his daily life, something to worship, something to beseech, something to blame. And what could provide all this better than the familiar and yet mysterious world of animals? So an enormous swing of attitude took us from hunting animals to worshipping them – not that we gave up hunting and then using them.

The first concrete forms of admiration, worship or passion are works of art, and the earliest are carvings. The most ancient are of the Venuses – fat females with large breasts and pudenda and the oldest was found in the Hohle Fels cave in South Germany and dates from the Aurignacian period some 35,000 years ago, an early forerunner of obesity! But the earliest animal found so far, is the Vogelherd Horse from 33,000BC. Carved out of mammoth ivory it is only 4.8cm long but the curved neck of a stallion shows an idealised power and breeding far beyond the sturdy ponies of the time – an inspired peep into the future. It may have been carried as a sort of charm to assist the hunting, for it is certainly worn smooth from being handled. Also in Swabia, in Germany was discovered the oldest anthropomorphic carving of a Lion Man, which points to some sort of early form of religion. Also dating from about 35,000 years ago are the Chauvet Caves in the Ardèche region of France. The walls are covered with over four hundred animal paintings of bears, hyenas, rhinos, mammoths and horses. The next oldest known site (25,000 years old) is at Pech-Merle near Cahors, also in France and possesses the most startling paintings of spotted horses. The Lascaux Caves are amongst the most famous but they date from Palaeolithic times only 16,000 years ago, but in the Hall of the Bulls there is a great tribute to the ancestor of our cattle – the Aurochs. There are also representations of men wearing reindeer headdresses; whether it was a form of worship, a way of celebrating the existence of these beautiful creatures, or whether it was a kind of hunting magic, we may never know.

Another beautiful carving from a mammoth tusk is of a pair of swimming reindeer from the late Magdelian age about 13,000 years ago. It is held in the British Museum.

Art, as we have seen, can be considered the greatest tribute of all to these special animals.

The proliferation of images of these six is uncountable – carvings, sculptures, paintings on walls and on canvas from one end of Europe to the other end of Asia. Some animals are crying for the attention of the artist far more than others and our particular six are amongst those. That is not to say we do not practice our arts on most of the other animals of our world, both the wild and the domesticated. Bast the cat goddess in Egypt two thousand years before Christ, has turned up in tombs, in numerous guises – bronze and stone and even gold. Sheep seem to have been greatly represented in heraldry and religion, and appear as the Lamb of God on many stained glass windows. Goats have made sporadic appearances, sometimes hybridised with men as satyrs and very memorably in the eccentric Picasso statue. The Chauvet Caves and their depictions of horses and cattle have already been mentioned. The Palaeolithic Altamira Caves in Northern Spain show goats and also horses, bison, reindeer and aurochs. Strangely enough the Lascaux caves are full of cattle, bison, horses and stags but show no reindeer, which must have been the staple diet of the artists, at the time.

Some of the earliest appearances of dogs in art were found when Howard Carter opened Tutankhamen's tomb and paintings of Anubis the dog god came to light. It was the start of a 'breed' that lived on as similar dogs were taken to Ibiza and became a 'breed' called Ibizan Hounds which look uncannily like Anubis.

Dogs appear very much in pictures of English aristocrats in the C17[th] and C18[th].

To mention Van Dyck again, he painted the five eldest children of Charles I in 1637 in which the future Charles II is shown patting an enormous dog. One of the most amusing if irreverent is the C17[th] painting by the Dutch painter Emanuel de Witte of a dog cocking his leg in a church. Throughout the Renaissance and Baroque periods dogs appear in every guise from hunting through to the religious and mythical. They are literally everywhere.

In the recently discovered tomb of Khnum-Hotep III (12th Dynasty) at Beni Hasan in Egypt there are pictures of cattle and hounds from C20[th] BC and even a pair of donkeys, male and female with two children riding on the back of the jack.

Horses, elephants and dogs are particularly depicted by the millions. Cattle figured more in early times, perhaps connected with worship. The Chinese produced lovely Tang statues of Camels but their Arab masters were not inclined to make images partly because their religion forbids representing the living image and partly because of their living conditions…not easy to carve or paint in the middle of a desert. However their distant cousin the Llama appears in a geoglyph (large design made on the ground, sometimes with stones), in the plains of Peru. The Nazca Lines, also geoglyphs, from 100BC – 700AD were made by removing the desert varnished rocks from light coloured soil in an area of desert 100 miles south of Lima. Gigantic depictions were made of a dog, as well as spiders, lizards and monkeys and the famous geometrical humming bird, and of course the llama.

Terracotta horses and chariots and dating back to more than 2,000 years ago were found buried in Pasryk, in central Asia. Other outstanding samples are the Mosaics in Carthage (Tunis) depicting the horses used by Hannibal (247 – 183 BC). One of the earliest written records of the Horse comes from a Babylonian tablet from about 1750BC.

Chinese art also drew on the "Heavenly Horses" for inspiration. Other often-reproduced examples are the Chinese T'ang pottery horses, (618 – 907 AD) which took the place of real horses being buried alive in the graves of noblemen so that their spirit form would be unmarred and their owners supplied with habitual transport in the after world. One of the Tang Emperors Hsuan-Tsuang owned no less than 40,000 horses, a constant inspiration for his famous court painter Han Kan.

Then to come down to more modern times the horses of the Parthenon are amongst the most world famous, and can still be easily seen in the British Museum. Were they as small in relation to their riders as they are represented or was it artistic licence? Probably the most legendary and larger than life statues are the four equestrian bronzes on St Marks in Venice. This ancient quadriga was the inspiration for the ones on the Brandenburg Gate in Berlin, the Wellington Arch in London and the Arc de Triomphe in Paris. Also in Italy, in Padua there stands the awe-inspiring statue of the *condottiere* Gattamelata created by Donatello in the mid C15[th]. Only when you see it 'in the flesh' do you realise that the legs are in pacing mode. Da Vinci and Rubens were the first European painters to portray horses in a lifelike style. William Hogarth (1697-1764) the painter, in his book "Analysis of Beauty" explores amongst other things the beauty and grace of a horse of noble breed.

In the British Museum, there hangs an engraving by Stubbs of the first zebra ever seen in England. It had been brought back from the Cape and presented to Queen Charlotte in 1762 and then kept in the Royal Menagerie.

For all these beautiful horses a record needed keeping and by wonderful luck there was the very man, George Stubbs the painter (1724-1806). He was possibly the greatest equestrian artist and painted such historic horses as Gimcrack, Eclipse and the Godolphin Arabian whose friend Grimalkin the cat died, after which the stallion who sired ninety foals, attacked any other cat he met. Probably his most famous painting is of Whistlejacket in 1762 for the 2nd Marquis of Rockingham. Apart from its size and total reality it is very unusual for the times in having a completely plain background. He also produced a series of drawings "The Anatomy of the Horse" in 1766 compiled from dissecting up and studying the anatomy from the muscles down until he reached the skeleton. The smell must have been something awful. The original drawings are in the Royal Academy. But it wasn't until Edward Muybridge with his ZoOpraxicope camera took a series of photographs of a horse in motion that it was found that a horse at full gallop takes all four legs off the ground. Before that they were shown galloping like rocking horses. Other memorable and famous equestrian paintings are Botticelli's painting of a centaur with Pallas Athene, in the Uffizi Gallery in Florence; Rembrandt's "The Polish Rider" of a C17th Polish light cavalryman, complete with bow and arrows, now in the Frick Museum in New York; Van Dyck's lovely equestrian painting of Charles I in the National Gallery, London. In both the last two portraits the horses are shown with long curling manes and more importantly hold their legs correctly and alternate.

Then in 1974 one of the most exciting discoveries of the art world came to light – the Terracotta army which Shi Huang Ti the 'Tiger of Qin' and the first Emperor of China had had made in 210BC. They were discovered by some local farmers digging a well and they then erupted onto the world stage. This Terracotta Army consisting of 7,000 soldiers, 600 horses and 100 chariots had remained hidden for 2,000 years and there are still unopened tombs. Every life-size soldier and horse has its very own character; there is nothing mass produced about them.

Elephants also figure largely in art. They are not only paintable but even more sculptural. Just to select a few examples there are cave paintings of mammoths dated to over 15,000 years ago at La Madeleine in the

Dordogne. There is also a famous cave painting – almost a cartoon in Spain at el Pindal of just the outline of an elephant wearing an enormous heart from possibly the Aurignacian period (30,000 years ago). The Sahara caves of Wadi Zerai and Tassili in Algeria show a tantalising glimpse of elephants living in the pre-desert conditions of the time. They give a sort of pictorial history of the arrival of various animals down the centuries in four chronological eras starting probably well before 4,500BC and of immense historic importance. The earliest demonstrates wild animals living in what was obviously rich grassland that existed before the desert we now have; the second shows cattle which only arrived there sometime between 4,500 and 4,000BC; in the third is displayed the first appearance of horses dating from about 2,000BC; and finally the advent of the camel which only arrived at the time of the birth of Christ.

Some of the smallest depictions of elephants from 3,000BC are among the Paki-Mohenjo-Daro seals carved in steatite from the cradle of elephant taming – the Indus valley. At the other end of the scale is the fabulous and famous frieze of elephants at Angkor Wat (started C9[th] AD), facing East, where every elephant is different in posture, movement and personality. But elephants don't just appear in art, female elephants have actually been taught to paint. Examples are Mamie an African elephant at Knoxville Zoo, Annabelle an Indian elephant at Anchorage in Alaska and Renée in Toledo Zoo in Ohio. Renée received formal art training in 1995 and to raise money to celebrate a Jazz pianist called Art Tatum is now painting a piano to be auctioned.

There are literally hundreds of elephants on Hindu temples in Udaipur, Jaisalmer and Deshnok in Rajasthan. The Sassanid (Persian) Kings had life size pictures carved into rocks depicting military events and elephants being ridden on a wild boar hunt. They even appear painted on the walls of modern houses in the Rajasthani villages.

But for real worship of the kind we can relate to, we must go forward to the earliest civilisations such as the Sumerians who from at least 3,500BC had an enormous and ever changing pantheon of gods and goddesses of whom some had a longer reign than others. Such a one was the sun-goddess Gula and her dog. In fact the alter egos of various goddesses of healing were dogs, such as Bau, who was originally one of the three main deities of the Sumerian pantheon. She is portrayed as a dog-headed goddess of healing and life. Dogs are her sacred animals and there were dog and puppy burials found at her temples. The belief that a dog's licking of wounds aided healing seems connected with Bau's position of goddess of physicians. One cylinder seal shows a healing ritual with a dog on the roof with a basin. Many figurines of dogs were found in Sumerian digs in S. Iraq, and many skeletons of buried dogs. One figurine instructed "Don't stop to think… bite!"

The Egyptian approach to dogs was unique; they worshiped a god-dog called Anubis, who had the head of a jackal, the body of a man and was entrusted with the guardianship of the dead. This was an extremely important post in the Egyptian pantheon because of their obsession with death. Their tombs, their pyramids and their ritual of embalming were all uniquely Egyptian. This last process of making mummies was also in the care of Anubis and not only people were embalmed but dogs, cats and even sacred ibises. Anubis also had the job of weighing the heart of the dead against the feather of truth. Perhaps the Anubis connection came from the jackals who scavenged round burial grounds looking for food and digging up the edible bones of corpses. So maybe the earliest reason for tombs was to protect the dead from the predations of jackals. Public processions were always led by a statue of Anubis carried out in front. Down the years he changed and was taken over by Osiris in Egypt but joined forces literally with the Greek god Hermes and became Hermanubis, a kind of mongrel dog god who was worshipped by the Romans right up until C2nd. Another Egyptian dog god was Wepwawet which meant "Opener of the Way" and he was identified with the town of Abydos, the home of Egypt's 1st Dynasty which makes him very old indeed.

The Greeks named two of the Egyptian towns Lycopolis meaning City of Wolves and the other dog town was Hardai in Upper Egypt named Cynopolis which meant City of the Dog by the Greeks and is teeming with dog cemeteries. There is a particularly good statue of Anubis in jackal form that was found in the tomb of Tutankhamen, and prayers to Anubis are found on many of the most ancient tombs. Set or Seth was a curious hybrid god which had the body of a greyhound but more of a deer's head and no one has managed to positively work it out, but when he wasn't being a man with red hair or a crocodile he was god of the desert and also, not surprisingly, of chaos.

Dogs were buried in family tombs and the members of the dog's family would shave their heads in mourning. Another tradition of the early Egyptians was to swear "By the Dog" and that signified an oath they would not break.

Moving on to the Greeks, Artemis was a goddess who could take on the appearance of a wolf. Homer described her "Artemis of the wildland, Mistress of Animals". She bore Zeus the twins Apollo and Artemis; the latter was known as the "Wolf Goddess" and had a wolf on her shield. Both she and Diana, her Roman counterpart, always had a pack of hounds with them, given by Pan. The wolf was also Apollo's sacred animal, perhaps in memory of his mother.

Hecate the triple goddess was always associated with dogs. And she had a special pet Cerberus, the three-headed dog who guarded the underworld, Hades, and prevented the dead who crossed the River Styx from escaping back into life – yet another dog god associated with death. He also figured as the most difficult of all the twelve labours of Hercules who had to capture him alive. Cerberus seemed to have had any number of heads ranging from 2 to 50 but 3 seems to have been the favourite number; the heads had manes of live serpents and he also sported a dragon's tail.

Dogs also featured in Roman myths and probably the best known is the she-wolf who suckled Romulus and Remus the founders of Rome. The she-wolf has remained honoured as the emblem of Rome.

The third oldest civilisation, that of the Indus Valley also celebrated a dog called Sarama in the Rigveda(1,700-1,100BC), which belonged to the all-important goddess Indra. Her name means the 'fleet one' and she was sent to retrieve some stolen cattle. Two of Sarama's sons were given the task of guarding the road of the dead, another example of the association of dogs with the dead. The Rigveda predates the Brahmin classification of dogs as unclean.

In Central Asia Genghis Khan considered himself to be descended from a wolf and called his generals 'The Dogs of War'. In ancient Mongolia dogs were as revered as parents. In modern times the Mongolians, who were stripped of their surnames during their communist revolution, have now largely adopted the name 'Borjigin' which means Bluewolf and was the name of the most renowned clan in that part of Asia in earlier times.

This close human/dog tradition came down from Roman times when there was a firm belief in the existence of people with the heads of dogs that were called Cynocephali. Even Herodotus (485-425BC) and Pliny the Elder (23-79AD) believed in them and metaphorically placed them in Ethiopia, which was a semi-mythical place itself and variously situated off Africa, near India or even in the Andaman Islands. These creatures barked instead of speaking and could spew forth flames. In mediaeval times the priests were worried about how to treat them. Did they have the souls of animals or were they descended from Adam? It was rumoured they lived in villages and covered their private parts. Two other unreliable witnesses Marco Polo and Sir John Mandeville also firmly credited their existence.

Dogs were not only gods but became firstly through hunting our closest companions and friends. However this close relationship between dog and man is not true for all mankind. The ancient Israelites were totally anti-dog and their lowest insults was to call someone a 'dead dog' or a 'dog's head'.

Nor do dogs get a very good press in the Bible. They are shown in many instances as outcast animals living on the streets and eating garbage. Although dogs were obviously useful and even essential as guard dogs and sheep dogs and for hunting they were mainly still regarded as unclean.

"Like a dog that returns to its vomit is a fool who reverts to his folly." Proverbs 26:11 [15]

"Outside are the dogs and sorcerers and the sexually immoral and murderers and idolaters, and everyone who loves and practices falsehood." Revelation 22:15

"Do not give dogs what is holy, and do not throw your pearls before pigs, lest they trample them underfoot and turn to attack you." Matthew 7

But probably the biggest group of those that believe dogs to be unclean are the Islamists. They believe dogs to be the most impure of all animals and see them as '*haraam*' forbidden, or '*kelb*' and have to wash an inconceivable number of times if they have been touched by them. Clearly Mohammed had a real problem with them and in the Hadith (sayings attributed to Muhammad) he ordered them killed, especially the black ones, which he equated with devils. So tainted are dogs seen to be that if a kitchen utensil is touched by one, it has to be washed seven times and rubbed with sand the eighth time. One good side effect is that in Islamic culture eating dogs is forbidden. Also it was strictly forbidden to make money out of selling a dog and this was placed on a par with making money out of prostitution, witchcraft or usury. The Muslim cultures of Iran and Iraq have rounded up their dogs and slaughtered them. During the Iraq War the Americans made full use of this horror of dogs and used them to intimidate Muslim prisoners.

However, even in the Muslim world, there were exceptions to the ban on dogs – it was permitted to keep dogs for the purposes of hunting or for guarding sheep. And the truly great exception, made by the Bedouins, was the Saluki. They not only carefully bred them for hunting but, went to the other extreme and considered them 'the gift of Allah' or 'the Noble One,' (al-Hor) or 'the Blessed One' (al-Baraha), and actually let them share their tents. They were usually carried around on their camels and horses until a quarry was sighted.

In Ashkelon, an ancient Canaanite city, thirty miles south of Tel Aviv, a huge dog cemetery has been recently excavated dating back to Persian times. Several hundred dogs, which seem to be of the greyhound type had been meticulously buried, with their tails tucked between their legs.

Perhaps the nation with the most variable attitudes towards dogs down the ages is China. They are mentioned on oracle bones from the Shang period (1,300 – 1028BC). During that period it was the custom to sacrifice dogs to bring good luck to the building of tombs and royal palaces and there were other rites in which dogs were dismembered and buried to placate the four winds or to keep away pestilence. One horrible ritual enacted was when the Emperor in a jade chariot crushed a dog under the wheels. But they did figure as one of the 12 signs of the Chinese zodiac and that made them popular as tomb figurines because people believed they would protect them from evil and help them enjoy a peaceful life in heaven. These beautifully sculpted little pottery dogs were discovered by archaeologists in Han Dynasty tombs (206BC-AD224), buried alongside their owners. But even then dogs were divided into three categories: watchdogs, hunting dogs and dogs for eating. There was another exception – as pet dogs for the aristocracy and the Emperor Ling Ti (156 – 189) actually gave royal titles to the palace dogs. The Imperial family even took to dog breeding themselves, probably mainly Pekinese or Foo dogs as they were called, and the Shih Tzu which acted as companions to the Imperial ladies and the palace eunuchs. These are early examples of 'paedomorphism' or the deliberate creation of a juvenile face by selective breeding.

Temples were usually guarded by a large pair of lion dog statues. But then during the Cultural Revolution (beginning 1966) there was another swing in the Chinese culture and dogs were considered bourgeois. This led to so many millions being killed that China's breeding stock was utterly depleted. Now with the advent of the one child family, people are electing to have a dog as a second child. But of course they have little idea of how to care for, feed, or exercise their new pets. In Hong Kong for example I heard that the servants were instructed to take the new pets for walks – carrying them in their arms! And in Shanghai there is talk of restricting dogs, like children, to only one to a family.

In Vietnam, dogs are bred for eating but even more awful are the gangs who go round stealing people's dogs which are guard dogs or working dogs or pets and sell them to the dog meat butchers. They are kept in unbelievably cruel conditions. The thieves are making a fortune and can be very violent if intervened but now some of the villages are fighting back and have killed thieves. It all escalated after the Soi Dog Foundation worked with the Thai Government stopped the trade from Thailand. Newly affluent Vietnam now splurges out on dog meat restaurants with their newly found wealth. Another incentive is that dog meat is meant to improve the male sex prowess.

Attitudes in Tibet, not surprisingly, are different and both the Dalai Lama and Panchen Lama had dogs for pets. Tibet has its own very special very enormous breed of dog – the Tibetan Mastiff. It is also an ancient breed and looks very lion like. Rich Chinese will pay anything up to half a million pounds for one.

In North America many of the Amerindians on the Pacific coast have the wolf carved on their totem poles, and killing a wolf for those clans is taboo. The Totem Poles were pillars made of cedar trees and depictions of animals that were part of the clan's kin were beautifully carved on them. These showed the animal spirits connected to the village ancestors. The animals were not worshipped as gods but were the equivalent of a mediaeval coat of arms. They also used totem poles as 'shame poles' to expose a debtor and were used for public ridicule.

Not surprisingly wolves come into the Norse mythology. The Norse king of the gods Wotan (Odin) who is so celebrated by Wagner in his Ring Cycle was accompanied everywhere by two wolves, Freki, which means 'ravenous' and Geri which means 'greedy'. Finally Wotan's death was an epic event in which he was killed by the gigantic wolf Fenrir.

Now for cattle – when they first entered the lives of Palaeolithic man, they came, not as workers but as deities. It was the greatest good fortune for them that this happened as otherwise they would probably have followed the only too many animals man has sent into extinction. The ancestral aurochs Cave paintings at Lascaux in France and Altimira in Spain from the Palaeolithic Period (17,000BC) show them in all their grandeur and perhaps as an early form of worship?

As it happened, Bulls became the masculine symbol of domination, the ultimate macho image and were worshipped all over Asia, the Middle East, and Africa. They continued in this role for thousands of years and in some places, such as India, cattle are still held in great reverence. Even in the Western civilisation this potent symbol is still celebrated as the Bull Market on the Stock Exchange, and the frantic value placed on beef and hamburgers in the Western world. The very word 'cattle' is derived from chattel and also capital, in the economic sense and came originally from the Latin word for head – 'caput'. So they became a visible, movable treasure, a sort of four-legged bank account.

The earliest worship of Bulls was introduced by the first Egyptian king Narmer-Menes (3,150BC) in order to unite all Egypt into an empire. This very first empire had the first really universal religion centred round the Bull god Apis, and his wife the cow goddess Hathor, who together ruled the skies. On earth a real bull was worshipped, kept for a year in the greatest of luxury, with a harem of nubile young cows, used for foretelling the future and then sacrificed, embalmed and entombed with as much ceremony as that afforded to royalty. These Apis bulls bred for religious reasons were black with white spots and a black spot on the tongue; and this cult continued for a thousand years. An Apis calf could be identified by certain distinct markings: the black calf had a white diamond on its forehead, an image of an eagle on its back, double the number of hairs on its tail, and a scarab mark under its tongue. Since the Apis was so sacred, it stands to reason that its mother (referred to as the "Isis cow") was revered as well.

In the Cairo Museum is kept an amazing bit of history – the Narmer Palette, carved out of siltstone and dating from C31st BC. It is 63cm long so was probably part of the temple regalia. Carved on it is the first king of Egypt Narmer, wearing a bull's tail hanging from his belt, and this remained part of the royal regalia throughout the early dynastic age. Also carved on it are four heads with horns, those of Bat the cow goddess of the sky and the Milky Way.

In Anatolia at Catal Hüyük the heads of Aurochs cattle, dating from 8,000 years ago were used to decorate the walls but the main worship seems to have been of a fat droopy breasted goddess – a sort of cow woman flanked by two lions.

Astarte, the Phoenician goddess to whom is ascribed the curious mixture of fertility, motherhood and war carried the horns of a cow. She was the equivalent of another cow goddess Ishtar of the Hittites. The first record of Astarte is dated 1,478BC but the cult was already ancient, dating from the Bronze and Neolithic ages. She was the extremely blood thirsty Mother of the Universe and seems to be the rather unlikely early prototype for the Virgin Mary, as has been suggested. The Phoenicians spread this very strong religion while voyaging across the Mediterranean to Greece, Italy and as far as Spain. The very name 'Italia' means 'Land of Cattle'.

The Minoans on Crete seemed to have been particularly involved with Bull religions and myths. Their King and Queen used to dress up as a Bull and a Cow and the walls of the Palace of Knossos were covered in the most beautiful frescoes depicting the Aurochs. The Minotaur, one of the most famous of mythical creations was the offspring of King Minos's wife Pasiphaë who had fallen in love with a white bull given to Minos by Poseidon. The bull was meant for sacrifice but it was so beautiful that Minos had kept it with the subsequent dreadful consequence. The wretched Minotaur was finally slaughtered in his labyrinth by Theseus, who went on to capture the Bull of Marathon. In yet another famous legend Europa, the princess of Tyre was abducted by Zeus disguised as a bull from the sea, who carrying her on his back, swam across to Crete, where Europa became the first Queen.

Another bull-centred religion taken up by the Romans in the first few centuries AD was Mithraism, just at the same time as the Christian religion arose. It was a very secretive religion and nothing was written to explain its origins. A 'temple' consisted of an underground passage, with a vaulted ceiling, dominated by a "tauroctony" or bull-slaying sculpture or painting. The subject would be of an athletic young man in a Greek tunic and Phrygian hat astride a crouching bull, clutching his muzzle with one hand and jabbing a sword into the bull's shoulder with the other. The extraordinary thing about these depictions was the dog and the snake drinking the bull's blood and the scorpion attacking its testicles. Maybe the religion was originally rooted in Zoroastrianism or maybe it was brought over by Celtic nomads. These temples existed wherever the Roman army went and there are quite a few underneath Rome itself. Other questions arise. Was it based on a Persian myth? Or was it connected to the signs of the Zodiac? Anyway whatever it was or wherever it came from, the Roman Catholic religion viewed it as pagan and outlawed it. This rite was apparently last performed for the Emperor Diocletian, but obviously forbidden by the Christian emperors.

During the reign of Nero there was a resurrection of a minor god Attis with an April festival celebrated with self-flagellating and piercing. Priests fully clad in togas were initiated by lying down beneath wooden slats while a sacred bull, decked out in flowers, was being sacrificed on the platform overhead. As the bull's blood streamed down through the slats onto the priest the devout members of the public confirmed their vow of celibacy by self-castration. The crowd cheered, and then burned the bull.

Turning to the Christian religion, there is a biblical story of the Golden Calf in Exodus 32:4 that probably arose from the concept of the Sacred Bull. While Moses was away on Mount Sinai receiving the Ten Commandments, the calf was an idol provided for the Israelites by Moses' brother Aaron who cleverly managed to manufacture it out of gold earrings brought from Egypt. When Moses returned after 40 days and nights he found the Lord waxing furious with the Israelites for worshipping the Golden Calf. It was particularly unfortunate as the second commandment on the tablets he had just brought back was against idolatry. Moses pleaded for his fellows and the Golden Calf had to be burned, ground up to powder, scattered on water and finally drunk by the Israelites.

In the Indo-Iranian culture the Bull-god ruled supreme, and the very title 'Gopa', given to a successful warlord meant Lord of Cattle.

In the Hindu religion the cow is a symbol of wealth and strength and is frequently mentioned in the Hindu scriptures dating back to a form called the Vedas, written in Sanskrit in the C2nd but derived from a much older oral tradition and the dependence on cows for all subsistence gave them the double status of caretaker/mother. They give the cow the same sacred status as the Roman Catholics give the Virgin Mary – the Mother of Life. One of the main gods Krishna was brought up in a family of cow herders, and given the name 'Govinda' which means Protector of the Cows. Also Shiva is traditionally said to ride on the back of a bull named Nandi who is a god in his own right. His origins were in the dairy orientated early Indus valley civilisation and there are temples in India built to him. There is one particular one in Basavanagudi near Bangalore built in 1537 that houses the most enormous granite statue of Nandi, complete with his Zebu hump, which stands 15 feet high and 20 feet long and has become blackened and shiny with centuries of rubbing with charcoal and oil.

Not only is the cow itself considered holy but everything that come out of it is sacred as well – their milk, curds, dung and butter will cleanse the body and purify the soul. Even their urine is considered sacred and used on sick children. The dust of footprints of cows also has religious meaning. Hindu livestock has entered the English language in the form of the expression of shock ("Holy cow!") and to describe something that is preserved at great length for no rational reason ("sacred cows"). Hindus believe that each of the 200 million odd cows in India contain 330 million gods and goddesses. Krishna, the god of mercy and childhood, was a cowherd and a divine charioteer. At festivals honoring Krishna, priests shape cow dung into images of Shiva, the god of revenge who rode through heaven on a bull named Nandi. Nandi's image marks the entrance to Shiva temples. Priests are called in to bless new-born calves. Kamadhenu the cow mother who grants all wishes is sometimes depicted all cow and sometimes with the body of a cow but the head and breasts of a woman. Cows wondering the streets are often ornamented with marigolds or beads as religious decoration.

Meat from cattle is only eaten by 'untouchables' and Muslims and other religions. Holy men smear their bodies with a mix of milk, curds, butter, urine and dung. No one is exactly sure when the custom of cow worship became widely practiced but the taboo on eating beef began in earnest around A.D. 500 when religious texts began associating it with the lowest castes. Some scholars have suggested custom may have coincided with the expansion of agriculture when cows became important ploughing animals. It has been suggested that the taboo was linked with beliefs about reincarnations and the sanctity of life of animals, particularly cows. In 1882 the All-Party Cow Protection movement was founded by Swami Dayananda Saraswati to prohibit the slaughter of cows in British India. The Muslims, partly in defiance to the Hindu sacrificed cows at their Bakr-Id festivals and this led to riots. More cow-related riots took place – fifteen alone between 1883 and 1891. And there was a bad one in Bihar in 1917. One of the reasons for the Indian Mutiny in 1857 was the belief that all new bullets were greased with pig and cow fat and as loading required biting the bullet; this meant being contaminated by an unclean and a holy animal.

Zoroastrianism, founded in about C6th BC, is related to Hinduism and also believes in protecting cows. One of their beliefs is that from the first bull's seed came all the plants in the world. With the advent of Islam into Persia/Iran Zoroastrianism was marginalised and moved largely to India and also the USA.

In early times, Oxen and cows were used as the most ancient form of currency. In the age of Homer the value of Achilles shield was calculated in oxen. 'The Armour of Diomedes,' said Homer 'cost only nine oxen; but that of Glaucus cost one hundred oxen.' Diomedes, an Achaean King was considered to be only second to Achilles, Glaucus was the Greek sea-god.

The Yak was not just the mainstay of everyday Tibet, just like with Bulls, another aspect of the Yak and its influence on man was Worship and the age-old connection to male fertility. It figured in a kind of shamanistic religion pre-dating Buddhism which only arrived in Tibet in the 600s. This religion called Bon or Bon-po, was a form of devil worship of a yak-headed god with lyre shaped horns called Yama.

Not only are the horns quite common as decorations on house roofs but they have another unique but rather grisly use. There is a caste of Tibetans called Ragyabas, the equivalent of the Untouchables of India. Some of them are employed to break up the bodies of the dead and pulverize the bones and flesh together and then feed the vultures at a funeral site. For this they use yak horns as tools.

Bulls have figured in many myths and legends. Other famous bulls feature in an early Irish epic (from at least 800 AD) written in the vernacular, called 'The Cattle Raid of Cooley' (The Tain.) In this story Queen Maeve was jealous of her husband King Aillil who owned the White Horned Bull of Connaught. Her army, after a terrific battle stole a Brown Bull from Cooley. The two bulls clashed in a fearsome battle. To quote "Their eyes blazed red in their heads like balls of fire. Their cheeks and their nostrils swelled like Blacksmith's bellows in a forge." Finally the Brown Bull won and ended up with bits of the White Bull in torn pieces around his horns and ears. Wherever bits of the White Bull fell to earth, so was the place named after it, for example his loin was dropped at Ath Mor thereafter called Athlone, and where the liver (tromm) fell was named Ath Troim and the ribs (cliathac) gave the name to Ath Cliath (Dublin).

There is a lovely legend concerning the founding of Durham Cathedral. Apparently the monks carrying the corpse of St Cuthbert and looking for a suitable burial place were led by a milkmaid who had lost her dun cow and found him resting on a certain spot. For some undeclared reason this augured well for the burial of the saint and the subsequent founding of the cathedral. Another saint – Luke had a bull as his symbol which may have led to him being the patron saint of butchers as well as doctors and artists. The Evangelist St Luke is often shown as an ox in Christian art.

The worship of Cattle eventually took hold in many parts of the world and went right down through the African tribes. For example the Masai of East Africa lived entirely off their cattle, mainly on their milk and their blood, and in their tradition believe that all the cows on earth belong to them.

Of all the animals that enjoyed the most tremendous sacred status the Horse is probably the most glamorous. The splendour and magic of them was celebrated in the Bible and the Koran. Horses' speed, their beauty, their strength attracted the greatest fan club on earth. To ride a horse gave you power, ennobled you as equestrians over the pedestrians. This attracted such envy that during revolutions, the French and the Russian, a major focus was to kill the superior 'vehicles' of their 'oppressors'. Perhaps this feeling was partly behind the hunting ban in England. It certainly has transformed into mobs indulging in smashing cars, in modern times.

Another widespread custom was of horses being buried with their riders. An important discovery was made by Schliemann (1822 – 1890) the amateur archaeologist who discovered Troy, of Horses and chariots in the shaft graves of Mycenae (1600 – 1400BC). Undoubtedly in the Old Palace period of Crete (2,000 – 1600BC) donkeys, horses and wagons were in use linking the Palace and the surrounding agricultural production. Graves with horse equipment appeared in E. and Central Europe dating from after 800BC. The Scythian Kings practiced the dire custom of being buried with the bodies of their strangled court around them. Then one year later a huge outer circle would be created with the bodies of fifty horses.

One of the best known of mythical creatures is the Centaur half horse and half man. Legend has it that they were descended with the Lapiths of Thessaly from the hero Ixion (son of Apollo) and his union with a cloud. More prosaically they were probably the reaction of the non-riding culture of the Minoan Aegean world when confronted by the first mounted cavalry. Lucretius in the C1ˢᵗBC points out only too pragmatically that horses are at their breeding best at three years old whereas humans are still babies. But maybe it was the absence of female centaurs, who only first appeared on a Macedonian mosaic in C4ᵗʰ BC, that held back the breeding! But the appeal of centaurs clung on into mediaeval times, for example there is a pair of them on a capital head in Moissac Cathedral dated C12th. Many fictional horses are as famous as the real ones and one with special significance to the British Army is Pegasus, Poseidon's son who was the winged horse on which Bellerophon rode to battle against the Chimera. This living myth was chosen as the insignia of all British Airborne Troops and the battle for Pegasus Bridge on D-Day was one of the more momentous of the 2ⁿᵈ World War. Appropriately from Shakespeare's 'Cymbeline' comes this very wish "Oh for a horse with wings!"

Neptune or Poseidon, who as god of horses as well as the sea had his own curious mounts the Hippocampi, half horses with fish's tails. A splendid example is the Trevi fountain in Rome showing Poseidon's chariot drawn by hippocampi.

One of the best known Greek legends was about the Mares of Glaucus, who were prevented from breeding by their owner, in order to keep all their energy for winning chariot races. However Aphrodite, the love goddess appalled by their sexless plight came to their rescue. She complained to Zeus, telling him Glaucus was feeding his mares on human flesh. Zeus gave Aphrodite the all clear so she led the mares out at night to drink from her sacred well and to eat a special herb called 'hippomanes'. The result was that at their next outing to the funeral games of Pelias, the mares bolted and overthrew their chariot. Glaucus became entangled in the reins and dragged along and the mares turned on him and ate him alive. His ghost called "Taraxippus" or Horse-scarer" still haunts the Isthmus of Corinth. The white Unicorn with its single horn has always been popular down the ages. But their mode of capture seems extremely underhand – they are drawn by temptation to put their heads in the lap of a seated virgin. But by far the fastest horse has to be Sleipnir the eight legged horse that belonged to Odin and could gallop so fast he flew without the need of wings. Another story is that Demeter turned herself into a horse to reject Poseidon's advances but he triumphed by turning himself into a stallion and fathering a foal on her with human speech.

Greek gods were often associated with horses. Ares (the god of war) rode in a chariot with four white horses. (White horses are the symbol of the highest purity). Demeter (the goddess of women, marriage and agriculture) is depicted with the head of a black mare and the priestesses of her temple were called foals. Pallas Athene, the female goddess of wisdom who incidentally travelled in a chariot drawn by four white horses inspired Prylis, son of Hermes to suggest ending the siege of Troy and entering the city by means of a wooden horse; (Odysseus claimed all the credit for this stratagem) It was built by Epeius out of fir planks with a trap door fitted in one flank and large letters on the other consecrating it to Athene 'In thankful anticipation of a safe return to their homes, the Greeks dedicate this offering to the goddess'. Odysseus persuaded the bravest to climb up the rope ladder that hung from the trapdoor. Next day the Trojans reported its appearance and that the Greeks had gone. There were great arguments as to what to do with it. King Priam's daughter Cassandra the prophetess, warned it contained armed men, but she was always doomed not to be believed by a curse put on her by a rejected Apollo. Laocöon warned "You fools! Never trust a Greek even if he brings gifts! But Priam, unable to resist owning this beautiful wooden horse, had it brought in with the well-known dire results.

Tacitus tells how white horses were used by Germanic tribes for divination, as they are associated with the Sun Chariot, white being the rarest colour for a horse. Horse worship also took place in England in the Iron and Bronze ages for example the elegant 3,000 year old White Horse of Uffington.

The Bible is also full of references to the Horse. "A whip for the horse, a halter for the donkey, and a rod for the backs of fools!" (Pr 26:3) (NIV) Another interesting mention of horses in the Bible is when Elijah runs faster than Ahab's horse and chariot all the way to Jezreel (1Ki 18:46, Pr 147:10-11). Then there was the story of Haman (Ester 6), who thought to glorify himself by riding a horse through the city, he must have been thoroughly put out when the king bestowed this honour on Mordecai instead.

Famously in the Bible (Revelations 6.) are featured the sinister Four Horses of the Apocalypse: The White Horse whose rider carried a bow which stood for conquest; the second horse was red and the rider brandished a sword and 'took peace from the earth'; the third horse was black and the rider balanced a pair of scales in his hand, standing for famine; but the most ominous fourth horse was pale 'and behold a pale horse: and his name that sat upon him was Death…' and this horse was the bringer of death by sword, famine, plague and the wild beasts of the earth.

According to Arab legend Allah created the horse from a handful of the South Wind. The Koran has many references to the glory of the horse and the Bedouin horseman who was a refined instrument of war, a wielder of death to all infidels; "By the snorting war steeds, which strike fire with their hoofs as they gallop to the raid at dawn and with a trail of dust split apart a massed army; man is ungrateful to his Lord! To this he himself shall bear witness" ("The Chargers" [100:1-7]). Mohammed's horse called Al Borak, which means Lightening was rumoured to possess a human head and after his master's death in 632 bore him up to heaven.

Horses not only appeared in religious books like the Bible and the Koran but were surrounded by a host of myths.

Of the Christian saints St James, the patron saint of Spain rides a white horse, as so does St George who is also patron saint of huntsmen.

The Battle of White Horse during the Korean War was one of the bloodiest battles with the White Horse Hill changing hands twenty-four times. In fairy tales knights often rode on a white charger. Nursery rhyme "Ride a cock horse to Banbury Cross" refers to Queen Elizabeth who was visiting Banbury and when her carriage wheel broke rode a white horse up the hill, or some think it may refer to a Celtic goddess named Rhiannon.

Mohammed transformed his followers from herdsmen into equestrian warriors and promised them paradise as a reward for being slain in battle, a sort of forerunner of the suicide bombers of today. Al-Borak was the white horse that Muhammed rode to Jannah.

In Hindu mythology there are numerous mentions of white horses and even one with seven heads.

Asses have, for a long time held a lowly position on the social scale of animals and were never mentioned in either the Iliad or the Odyssey. They reached the Mediterranean countries 4,000 years ago but only arrived in Northern Europe in Mediaeval times. A mediaeval form of punishment was to make the victim ride naked on a donkey through the streets, facing backwards to endure taunting and humiliation. Two antipopes were so treated. However, on the cosmetic side of things, their milk was famous and baths were taken in it. And it was also taken as a medicine. A further use was of their skin for parchment. They were for many years the mainstay of the Mediterranean and got many mentions in the Bible including its most famous journey of all – the carrying of Jesus Christ into Jerusalem.

The ass was the symbol of two ancient gods – the Egyptian god Seth and the Greek god Dionysus. And Greek mythology contains the story of King Midas who was judged against Apollo in favour of Pan at a musical contest and Apollo changed his ears into a donkey's as a punishment. 'Donkey' has long been a term of opprobrium connected with stupidity and ignorance.

The camel in religion? They certainly march in trains through the lands of the three Monotheist religions, through the bible and through the Koran. The people of the deserts were entirely dependent on them but they do not, like our other five animals, seem to have been the object of worship. They were, together with horses and pigs, on the list of animals not clean to eat. They certainly show their well-known obstinacy by refusing to go through the eye of a needle!

There is an exception founded on the Zoroasterian religion of 6th BC. In Nietzche's "Thus Spoke Zarathustra" he describes the human spirit making progress towards the 'overman' and proceeding through three stages. First is the stage of the camel, where we renounce comfort and learn to discipline ourselves harshly. Second is the stage of the lion, where we defiantly assert our independence. Third is the stage of the child, where we find a new innocence and creativity. Achieving this stage is like reaching the summit of a mountain.

So not much about the camel in religion, however with their remote cousins the guanacos, there was a totally different attitude. In Peru the Incas had the only pack animal that had been domesticated outside Eurasia – the Llama which doubled up as creatures to be used for religious ceremonies and sacrifice. Even one of the Inca deities Urcuchillay was represented as a multi-coloured llama. But the main goddess was Pachamamma, the earth mother and her consort Inti, who were benevolent beings that inhabited the Andes and controlled volcanoes and earthquakes.

After the Spanish conquest during which they did their best to destroy the local religion, the Virgin Mary half took over from Pachamamma but not entirely. Even now after five hundred years of Roman Catholicism pilgrims still go to Pachacamac, which was a place of worship from very ancient times and bury offerings, such as corn and coca leaves there and also sacrifice chickens and even llamas to her.

Another god was Viracocha who had created all the animals and man from the mud of Lake Titicaca. The Catholic view of the Incas was that they were one of the 12 lost tribes of Israel and that the llamas had been one of the creatures in from the ark. The excuse the Spanish hid behind for their incredible brutality was that all non-Christians were creatures of the devil. When the Inca King threw away the, to him, unreadable bible, a massacre resulted by the affronted Spaniards. Tragically for history, as well as the horrendous killing, the Inca form of writing which was done with knotted cords and the Mayan form of hieroglyphics were lost forever.

In Inca times hunting was totally forbidden except for once a year when thousands of men took part in the Battu. This was partly a ceremony and partly a necessary cull and culminated in the annual meat meal. About one third of the Inca Empire's llamas and alpacas were reserved for royalty, the retinues of priests and for sacrifice. For the Royal family special garments were woven from the wool in the temples by special priests. A llama, preferably a white one was sacrificed every morning. While its head was turned to face the sun its throat was cut. Auguries were read from the entrails and the morning news was told to the Inca. Slaughter was undertaken by a butcher of the oldest or weakest animals, by cutting into the chest and wrenching out the aorta which causes no bleeding or suffering, unlike the terribly cruel halal method of slaughtering animals.

José de Acosta (1539-1600), a Jesuit priest tried to provide an explanation from the Holy Scriptures for the presence of *camelids* in the Andes. He is known as the Pliny of the New World and he also tried to supply explanations for earthquakes, volcanoes and tides. He also attempted to give reasons for altitude sickness and the treatment with coca leaves. He almost hints at evolution and his explanation of the llama family was "God wished to combine sheep and mules in a single species in order to ease the existence of the people who live in the sierras".

Our fifth animal, the Reindeer, figures in the Shamanism religion of the Tsaatan or 'Reindeer People' of NW Mongolia in Khuvsgul Province. Their lifestyle entirely centres round their reindeer as unlike some other nomadic tribes they do not have cattle, goats or sheep. Shamanism is a form of nature worship and is described as a 'technique of religious ecstasy.' It has the longest recorded history in the world from at least the Shang dynasty (ca. 1600–1046BC).

In N. Mongolia there is a sacred alliance between the Dukha reindeer nomads migrating through Mongolia's Hovsgol Provence's forests, which are considered to be inhabited by spirits of the ancestors.

As noted before, painting of animals, especially in pre-history could indicate a form of worship. And also mentioned have been the Chauvet Caves with their depictions of reindeer which were then living and being hunted right down into France. Father Christmas's sleigh has been famously pulled by reindeer ever since they figured in a poem dating back only to 1823 called 'A Visit from St Nicholas'. However as Christmas falls in mid-winter, unfortunately the only reindeer that would have retained their antlers at that time of year would be female and neutered males so some of the names like Prancer, Dancer, Dasher, Vixen are all right and Donner and Blitzen and Comet could refer to either sex, but the masculine names Rudolph and Cupid will have to go! Another absurd prohibition by Americans was practiced on a Swedish astronaut, Christer Fuglesang who was not allowed to bring dried reindeer with him on-board a shuttle mission. The Americans thought it a terrible thing to do just before Christmas. He had to take dried moose meat instead.

Elephants are a frequent subject of myth, fable, art and magic. Sanskrit texts originating in the Indus valley date back to over 1,500BC. One of the earliest mentions of elephants is in the Rig Veda, one of four Hindu canonical sacred texts from 1700 –1100BC. The word Veda means 'knowledge' in Sanskrit. Early gifts of elephants are recorded and one description is of an unwieldy present of 10,000 elephants and the same number of slave girls. Applicable to this story is the sardonic Thai curse "Give your enemy an elephant – and saddle him with the upkeep!"

Probably the best known and loved of the Hindu gods is Ganesh the "Lord of Beginnings", the god of wisdom and good luck, with the body of a man with a portly belly but an elephant's head and sometimes four arms. Later he became the patron saint of writers and merchants. Legend has it that the Lord Ganesh broke off one of his tusks for a pen and continued to write The Mahabharata until the bard Vyasa stopped him. He appears in literally hundreds of sculptures. The carving in Hampi Andra Pradesh is fifteen feet high and was hewn from a granite cliff in the C16[th.] Rather perversely the vehicle or *vahana* for carrying Ganesh, is a large rat. Another Hindu god is Indra, a god of the weather. His weapons are thunderbolts, and he is also a bringer of rains. Indra's particular *vahana* is a large white elephant called Airavata, often depicted with four tusks.

During the Mauryan Empire (600 – 321BC) the Pali Jataka stories were written. They concern Buddha's former births and are great folklore literature rather on the lines of Aesop's Fables. Here are stories of elephant festivals and marvellous tales of elephants and kings. Buddha or Bodhisatta figures largely and is re-incarnated several times as a white king-elephant. White again is shown as a special colour.

There has always been more mythology amongst the agricultural peoples of Asia than with the hunter-gatherers. In the rural villages of Asia elephants are considered cousins of the clouds – not so ridiculous when you relate their size and grey colouring and thunderous trumpeting! Thus they were also linked to the rainfall. And in Sumatra it was believed elephants caused lightening. These beliefs are no more bizarre than the belief, in mediaeval Europe, that elephants had no knees so had to sleep standing up leaning against a rock or tree.

In Tibet the elephant is held to be the holy symbol of Buddha. And in Siam/Thailand one of the Thai king's titles is 'King of the White Elephant'. These 'white' elephants were the object of worship or veneration in India, the very opposite of the West's interpretation as referring to an unwanted and awkward object. They are actually only semi-albino with a sort of pale salmon and freckled grey skin but they are painted white for ceremonies. In mediaeval Burma the white elephant was supposed to contain the soul of a departed king and likewise worshipped. One of the more bizarre sculptures of an elephant appears on Notre Dame Cathedral in Paris – with claws. One of the greatest elephant friezes in the world is at Ankhor Wat, each of the four sides of the temple is nearly half a mile and the bas relief of the dozens of elephants on the one facing the sunrise show amazing character and life, when you think they were sculpted in the early C12th. Elephants have great appeal and are carved in wood and ivory and sold in the Far East and Africa by the million; paintings of them too are popular.

Another amazing mythical kind of multi-animal of southern India is the Yaali, which has managed to merge bullhorns onto a lion's head with a crocodile snout ending in an elephant trunk – all carried on the body of a rampant horse. It is a very essential part of Pallava architecture from C6th to C10th and decorates the Palace of Malibaliporam.

Mores and attitudes towards animals present their own difficulties. The customs over the eating of the six special animals are complicated and changeable. One man's god is another man's dinner. Neolithic man probably ate any bit of meat he could get his hands on, even wolves, but once the famous partnership had struck up the eating of dogs definitely became taboo in certain civilizations. The Chinese with their multi-view on dogs have always eaten some even while they were making pets or guards from others. In the present day for example, they are importing St Bernard dogs for the very reason they breed large litters which grow and put on weight so quickly that at about 4-6 months they are ready for the table. And dog meat in China has a long history and was often referred to as 'fragrant meat' or 'mutton of the earth'. It is said that 13-16 millions dogs are consumed in Asia every year.

The Chinese are not the only consumers of dog flesh. The Aztecs raised special hairless dogs called xoloitzcuintles for the pot. They had the excuse that apart from dogs they only had turkeys as domestic creatures. Captain Cook found many places in the South Seas where dogs were being raised on a vegetarian diet as a preparation to be eaten. The Germans when on the point of starvations in their various wars consumed dogs under the euphemistic title of 'blockade mutton'. The Swiss, very surprisingly, eat dog in a couple of their cantons.

Not surprisingly, Islam bans the consumption of dog meat, and the Hindus and the Buddhists also regard it as totally unclean. Vegetarians must make better eating than carnivores so the South Sea Islanders were very sensible with their dog diet. Very few people seem to be neutral where dogs are concerned.

The next animal, the cow, also has its violently 'fors' and 'againsts'. In the Hindu world, slaughtering cows is considered 'anyaya' which means 'illogical' as cows are more useful alive than dead. This was not always the case, and in times of shortage, people slaughtered cattle to provide themselves with immediate food but for the long term it was a disaster and the only way to prohibit the habit was to attach a beef taboo to the Hindu religion. Some Chinese farmers also felt it short sighted to eat the animals that sustained their agriculture and certainly the Buddhists are against it, but more from the vegetarian point of view. Japan only lifted a ban on eating meat in 1882, soon after the Meiji Restoration and the end of the Shogunate. Christianity on the other hand was all for the eating of meat; but it had to be given up for Lent and on Fridays as a symbol of renunciation.

The Romans were so greedy about beef which no doubt was leading to obesity, thus causing dietary laws to be brought in. Eating beef was re-introduced into England by the Normans after their conquest and it also took off there in a major way. Somehow raw beef had become associated with power and male domination. In the Middle Ages gluttony was extremely common amongst nobility and was seen as a sign of well-being, and yet in Dante's inferno 1 of the 12 circles of Hell was especially reserved for gluttons; and the church frowned on it as a venial sin. Beef eating is still associated with having 'arrived' economically, and the upwardly mobile nations such as the Chinese are striving for beef on the table. Early examples of fat being considered beautiful are the female models, used by Rubens, who weighed over 200lbs.

In the modern USA the "age of Corpulence" has arrived with a vengeance, with hamburgers and other beef products being sold by huge chains of fast-food restaurants spreading their tentacles round the world but especially all over the USA. It is now estimated that there are over one billion people gorging and purging and obese. Balance that against a billion who are starving or mal nourished. That leaves roughly 4½ billion who are somewhere between overfed and starving.

Hippophagy means eating horse flesh and it has been practiced on and off since Man and horse first met. As we have seen those horses that had not left North America across the Bering Land Bridge suffered this fate and became extinct there in 10,400BC. The same would have happened in Eastern Asia if Man had not found better uses for horses than just as meat. Once chariots and riding took place horses were elevated to a near sacred status and it became unthinkable to consume the creatures that made mobility so easy. Their owners still ate horses when they grew too old to work and this practice continued down the centuries.

At some stage in Nordic mythology eating horseflesh became associated with Odin, which may have been one reason that Christianity banned it. Pope Gregory III in AD732 forbade eating horsemeat as he connected it with these pagan practices.

There were other often desperate reasons for eating horsemeat; one example was during the French Revolution when the killing of horses was linked to bringing down the aristocracy literally to the level of the proletariat. Certainly on the 1812 retreat from Moscow Napoleon's starving soldiers were surviving by cutting chunks out of living horses before they froze solid at temperatures of -40°.

Generally eating animals in the category of pets such as dogs, cats or horses is taboo but when the alternative is starvation then the taboo is broken down. The Battle of Tours in 732 AD showed the emergent importance of cavalry, so Gregory III made an effort to stop the practice of horse eating, and wrote to St Boniface and told him "It is a filthy and abominable custom". Horses were essential to stop the Muslim cavalry, which was threatening the Christian ascendant in Europe. His edicts are based on the same scripture as the Jewish prohibitions and this ban remained until the 18th century. It is forbidden by Jewish law, because the horse is not a ruminant, nor does it have cloven hooves. In Islamic law, horses are generally considered *makrah*, i.e. the meat is not *haram* (forbidden) but the eating of it is strongly discouraged. It is obviously forbidden in Hinduism as they practice vegetarianism.

It was the French who broke the taboo on eating horse flesh much to the horror of the English, and made it legal in 1866, when it was considered a 'basse viande' for the working class.

Consuming horseflesh is also part of the cuisine of countries as widespread as Japan and Kazakhstan. The Christianization of Iceland in 1000AD was achieved only when the Church promised that the Icelanders could continue to eat horsemeat; but once the Church had consolidated its power, the allowance was discontinued. In some parts of Canada and the USA horse meat is legal but not everywhere. This ban was lifted during WW II, as beef was very expensive and kept mainly to feed the troops. In the UK, this taboo is very strong indeed and horsemeat is banned even from commercial pet food. Horsemeat is also avoided by most people from the Balkans, mostly for ethical reasons, as the horse is considered to be a noble animal, or because eating horsemeat is associated with war time famine. A similar taboo exists in Poland.

Now to look at attitudes towards the camel – it is forbidden to eat camel meat in Leviticus 11:2-4 and in Deuteronomy 14:6-7; although the camel fulfils the criteria of having split feet and being a cud-chewer, the Torah still considered it "unclean". The eating of camel on the other hand is permitted in Islam, and indeed is traditional in Saudi Arabia and generally in the Arab world. The hump particularly is considered a delicacy and eaten on special occasions. Camel meat is for sale all over London and some of the Midlands towns. The eating of camel meat could be introduced in Australia where a million feral camels roam doing environmental damage on a grand scale. As for the camel's South American cousins, in Inca times the hunting of the Guanaco or Llamas was totally forbidden except for once a year when thousands of men took part in the Battu, a kind of hybrid hunt/race and for a brief moment there was Guanaco meat everywhere.

Now for elephant meat which has its attractions – an elephant is a vegetarian and has a large amount of meat. Judaism prohibits consumption of elephant meat, a taboo similar to the prohibition on camel meat. Islam also forbids consumption of elephant meat since the elephant is an herbivore with tusks. However, in Central and West Africa, elephants are hunted for their meat. Some people in Thailand also believe that eating elephant meat improves their sex lives and elephants are sometimes hunted specifically for this.

Islam has its own method of slaughter called Dhabiha, both the throat and the carotid arteries are cut, but not the spinal cord. The animal is then bled to death. The two important words are Halal meaning permissible and Haraam which signifies that the food is forbidden. Both McDonald's and KFC have managed to open halal restaurants in Sri Lanka.

In Israel the term Kosher means permissible and involves ritually slaughtered beef, sheep, goats and deer. Stunning the animal first is not permitted. Again the animal has to be bled clean, as consuming blood is forbidden. In the Old Testament, Leviticus 11: 1-47 is a list of animals prohibited for eating. It is very detailed about which animals have cloven feet and which chew the cud.

Our final animal caribou or reindeer is freely consumed in Alaska, Canada, Russia, Norway and Sweden and is very popular as a dish, but people in England and America are squeamish about the idea of eating reindeer meat, probably because of the Father Christmas story. It is however kosher and is available to Jews in Israel, and USA.

As we have seen our six mammals had a serious influence over our various religions and our diet – the second tradition being very much influenced by the first – our gods dictating our dinners. However as we shall see, there are happier, colourful and more exciting uses to which they can be put as is shown in the next two chapters.

ET CETERA

1. The Tamils in the C1st to the C3rd AD were so closely connected to elephants that they had 44 names for elephant species, four separate ones for females and even five more for the calves. They also had separate names for the body parts of the elephant.

2. In Switzerland in 1577 some giant bones were dug up. They were assumed to be human and the Swiss church wanted to give them a Christian burial and a doctor at the time calculated them to belong to a 20ft giant, possibly a Cyclops because of the hole in the centre of the skull. Not until 200 years later were they were declared to belong to a mammoth, and of course the hole was where the trunk had been joined on.

---oo0Ooo---

Chapter 11

GAMES AND GLAMOUR

Games and Glamour Polo match. Bull fight

Charles Darwin pointed out that Nature was not perfect harmonious stasis but rather a turbulent engine of adaptation. Tennyson's view has always fought on, and always will.

> *"Who trusted God was love indeed*
> *And love Creation's final law*
> *Tho' Nature, red in tooth and claw*
> *With ravine, shriek'd against his creed"*
> *Alfred Tennyson*

> *"With monstrous gape sepulchral whales devour*
> *Shoals at a gulp, a million in an hour.*
> *-Air, earth and ocean, to astonish'd day*
> *One scene of blood, one mighty tomb display!*
> *From Hunger's arm the shafts of death are hurl'd*
> *And one great Slaughter-house the warring world!*
> *Erasmus Darwin*

Games *must* have started when wolf/dog puppies arrived in a cave with children! You can't imagine them not playing with each other. Then one thing must have led to another. Dogs have an inbuilt talent for retrieving things – later encouraged by selective breeding. It could have started with early hunting – a mixture of games and work. Now retrievers and other dogs bring back birds shot by their owners with the greatest care not to damage them and quite a few dogs like to play a private game with their owners who throw a ball for them to retrieve. In fact this is so popular with dog walkers who then exercise their dogs more than themselves that a special ball thrower has been designed.

One of the earliest known forms of games is a totally wild form of polo played in Afghanistan called Buskashi. This game, from the time of Genghiz Khan, was played in Central Asia by highly skilled horsemen called *chapandaz* with a dead goat or calf or sometimes with the severed head of a foe, which is snatched from player to player. It was and is the ultimate war game. For a start there are few rules and there are no boundaries of time or place. There is dust, there is sweat, and there is screaming excitement and the thunder of hooves. Raging adrenaline, danger, injury and death – all the ingredients of battle in those days, are there. The Taliban have banned Buzkashi in Afghanistan on the grounds of immorality, their word for having fun. In the more primitive form of the game called Tudabarai, there are no rules or even formal teams or even proper boundaries and there can be hundreds of contestants. In the more regulated form of the game called Qarajai, there are two teams of 10 – 12 riders, the winner gallops round a flag, carrying the goat which he throws into a marked circle called the "Circle of Justice".

Polo or Chogun, a far more ordered game, originated in Persia even before Darius the Great (485 BC.) and was probably named after the Tibetan word for ball 'pulu'. The first recorded match took place in 600BC between the Turkomans and the Persians. The Turkomans won! And unlike the savage Buzkashi, its popularity has taken it all round the world. It was a very popular game in China during the Han Dynasty, at the beginning of the C3[rd] AD. Amongst the most interesting of *mingqi* sculptures are of women playing polo, their polo sticks look more like hockey sticks.

The tea planters of India adopted the game and inevitably, as is the British way, founded clubs. The Calcutta Club founded in 1862 is the oldest existing polo club. The division of the game into chukkas is an Indian word. It was taken to Malta by returning planters. In England the most famous centres are Hurlingham, Windsor and Cowdray. The Argentine is famous for its players who often have a handicap of 8 or more. It is played in South Africa, Australia where there are 50 clubs, and New Zealand. Now over seventy countries round the world play the game. An extraordinary departure from polo on turf is polo on snow over ice played in St Moritz, Klosters and Kitzbühel in the Alps and Aspen in the USA with a large red ball obviously unlike the little white painted wooden ball of regular polo.

The highest polo ground in the world at 3,700m is at Shandar Top known as the Roof of the World. It is definitely a freestyle version of polo and there are No Rules. Health and Safety would have a succession of fits as the players don't all wear helmets or even boots. In the mountain area of Chitral on the Afghan border there is a game festival in which they also play polo but that is now coming under terrorist threats from the Taliban, like kite flying or any other form of enjoyment.

Polo, in China, was outlawed by the communist regime in the sixties, partly because they were anti-horse. But now Polo in China has started up again. Beijing property magnate Xia Yang, inspired by seeing on TV Prince Charles playing in a match v Sultan of Brunei started the first polo since before Mao's Cultural Revolution killed it off. There is no shortage of potential members of the new Beijing Sunny Time Polo Club founded in 2004 and which is hoping to attract some of China's 101 billionaires and 320,000 millionaires. They have lined the Club walls with paintings of uniformed hussars and hunting scenes from Britain! A good Polo pony can cost £90,000. Another club was founded in 2007 in Shanghai, called The 9 Dragons Hill Polo Club which is teamed up with a golf club and spa.

Yak Polo or Sarlagan Polo was started in the early 2000s as a tourist attraction but it has taken off in a big way. The highest polo ground in the world at 13,000ft is at the Boroghil Pass on the Afghan border which is on the ancient Silk Route.

An even more amazing form of polo is with elephants, which was started by a chance conversation in 1982 and there is now the WEPA (World Elephant Polo Association). Its headquarters is at Tiger Tops in the Royal Chitwan National Park, Nepal. There is a tournament held every year in Megauly and teams from Nepal and Thailand compete. There are a few bizarre rules such as forbidding an elephant to lie down in front of a goal mouth, nor may it pick up a ball with its trunk. The mallets are 2 meters long and ladies may use both hands. The elephant carries both a driver and a player and seems to enjoy the game. They changed from using a soccer ball, which the elephants liked to squash and pop, to the customary wooden polo ball.

There is even polo played from camels, both Bactrians and dromedaries, in Mongolia and Rajasthan but Dubai has really taken it on a big way and the games are organized by the Dubai Polo and Equestrian Club.

Another very early and exciting use of horses was horse racing, the origins of which lie in Central Asia. Racing can be viewed as another form of game. Racing competitively began amongst these diverse, often nomadic populations over 6,000 years ago. The first use of Chariots was being used for war but they also figured in games, for example the Greeks drove chariots as a national sport for over 1,500 years, a lot longer than any modern sport has existed. It was every horse owner's dream to take their horse to the Olympic Games. Horse racing had begun in the Olympics in 648BC, and was even more hazardous than nowadays as neither saddles nor stirrups had as yet been invented. One of the earliest known descriptions of a chariot race comes from Homer's 'Iliad' written in 750BC about the Trojan War. In that celebrated race there were five contestants who had to race their chariots to a goal and back. The race was judged by no less than Achilles, he who was later slain and dragged in dust behind a chariot circling the walls of Troy, but nothing to do with the race.

At the beginning the races were part of religious festivals called 'ludi'. These took place more and more frequently until they were being held 135 days a year. The Romans had a passion for chariot racing similar to the football mania of today. Only wealthy people could afford to pay for the training, equipment, and feeding of the charioteer and the horses. As a result, it was the owner who received the olive wreath of victory instead of the driver. But the best drivers were feted and grew very rich. The fame of one charioteer Gaius Appuleis Diocles has come down the centuries, as he won over a thousand races. He was the Stirling Moss or Gordon Richards of his day but amassed a far greater fortune, estimated to have been the equivalent of £9.6 billion, which makes our present celebrity earnings look like small beer. There were four teams the Reds, the Blues, the Whites and the Greens, Diocles started with the Whites but was not faithful to any of the teams and changed his allegiance from time to time. The races were held in the Circus Maximus in Rome which could hold a crowd of 150,000 and was 620m long. The chariots were 'quadrigas' meaning drawn by four horses and the races consisted of seven laps of the circuit round often very dangerous bends. The terrible and often fatal crashes that often occurred were called 'naufragia', which means shipwrecks. A deliciously embarrassing episode happened to the Emperor Nero who in 67AD drove a 10 horse chariot on one occasion and fell out of it. The quadriga is immortalised in statues on the Brandenburg Gate in Berlin, the Wellington Arch in London, the Bolshoi Theatre Russia and also on the Arc de Triomphe in Paris. Another famous racetrack was in the Hippodrome in Constantinople (Istanbul). This arena had a tunnel from the nearby palace to the Emperor's box used on one occasion to escape a baying mob.

There were also 2-horse chariot races, with separate races for chariots drawn by foals. Other races were between carts drawn by a team of 2 mules. The course was 12 laps around the stadium track (9 miles), and there were separate races for full-grown horses and foals, but the quadriga was the race that appealed the most.

Mounted horse racing extended across the empire and became a popular event right across the Empire's provinces and in AD 210, the Roman Empire brought horseracing to Britain. At a Roman encampment in Wetherby, Yorkshire (where a racecourse can still be found today) local horses were matched against Arabian horses brought over by the Roman Emperor Severus Septimus. This was the fatal beginning of English racing.

By the middle-ages, English knights themselves were bringing back horses to Britain from the Arabian Peninsular; horses that they had 'acquired' during the Crusades. With their introduction the origins of horseracing as we know it today were born. The descendants of these sires were bred and crossbred to create a horse that was very fast, yet strong. Almost all of the selective breeding was for one purpose, to produce the fastest horse on the track. A wide girth for a large lung capacity was essential and of course strong legs for hard running. The shoulder is long and sloped to allow a greater stride. The hind leg is long so that it can gain greater ground quickly. Everything about the breed suggests speed.

Racing had a rather checkered career in England. The kings of England were on the whole keen race promoters. Henry II held one of the earliest races at Smithfield in 1174 and King John (r 1199 – 1216) imported Eastern horses for a royal stud at Eltham. Richard I (r. 1189 – 1199) had knights racing their stallions for a £40 purse of gold but the first recorded race meeting was at Chester in 1512. Then under James I racing really became popular and this led to the start of pedigrees; he also developed the town that eventually became the headquarters of racing – Newmarket. Banned by the spoilsport Cromwell (1599-1658) who was, of course, with his Puritan ideals and anti-royalist views, dead against racing. (One can't help comparing him to the Taliban). Kings and Queens have been breeding racehorses for 500 years. One of greatest was Aureole bred by George VI and raced by the Queen. The present Queen is also mad keen on horses and racing and has won many races but not yet the coveted Derby. She also has a stud at Sandringham.

Then fortunes swung the other way – the restored monarch Charles II (1630-1685) seized Cromwell's seven best horses and under him, racing which was his favourite pastime, earned the soubriquet 'The Sport of Kings'. Due to his marriage to Catherine of Braganza (1638-1705) England acquired as part of her dowry Tangier and Bombay. So Charles was able to improve his breeding stock with Arab mares – called the 'royal mares'. In 1671 the winner of the Newmarket Plate was King Charles himself, the only English king to have actually ridden as a jockey and won a race; but how pushy was the rivalry of the rest of the field is something lost to antiquity! The horse he rode was called Old Roley, the nick name often applied to his royal owner. His mistress Nell Gwynn was reputed to be a heavy gambler. This penchant for gambling grew and grew into the C18[th] and applied to all classes and all mediums – dice, cards, cock fighting and most of all horse racing. The C18[th] also saw the beginning of organized racing at Epsom Downs. The first race meeting at was held at Goodwood in 1802, and racing on the downs has continued ever since

Horse racing has increased in popularity down the centuries and the descendants of the 3 famous Arabians the Darley, the Godolphin and the Byerly Turk are still the best. Most of us are hard pressed to trace our family tree beyond our great grandparents, so it's astonishing to realize the direct bloodline of a thoroughbred horse can be followed back through tens of generations.

Nowadays the most powerful forces in racing are the Coolmore Stud in Ireland and then from the 1970's back came the Arabian influence in the horse world in the shape of Sheikh Mohammed bin Rashid Al Maktoun, the ruler of Dubai and flush with oil money who began his own breeding and training stables – the Darley Stud. Early in 2010 he opened the Meydan Race Course which must be the most luxurious course in the world with a grandstand for 60,000 people and a five star hotel trackside. Gambling for Muslims is, of course, illegal so that must take away much of the point of racing. It may be against Sharia law but that doesn't mean gambling doesn't take place as has been seen recently in the Pakistani cricket team.

There are over 5,000 thoroughbred foals born in Great Britain every year, and over 110,000 born worldwide. Each one can trace their ancestry back through the father's line to one of these three stallions.

Racing today is a far cry from the informal events of antiquity. For the race goers there are extremely sophisticated grandstands, with bars and restaurants. Then there is the most advanced camera technology, but the main attraction is the thrill of seeing beautiful horses racing each other and the added excitement of having a chance of choosing the winner and making money. Unfortunately some people have been financially ruined by their lack of judgment.

By the time of Queen Anne's reign (1702-1714), match-racing just between two horses was gradually being superseded by races involving larger numbers of horses and greater numbers of spectators placing money on the outcome.

The Jockey Club was started about 1750 and Sir Charles Bunbury was the first head. It began when a group of racing gentlemen used to meet regularly at the Star and Garter, a pub in St James's. It was neither a club nor had jockeys for members and it soon moved to Newmarket. It devised rules to ensure that racing was run fairly on Newmarket Heath then gradually their practices were adopted all over the country. A very closed and elite membership ranged from Royal Dukes to members of parliament. Bunbury and Lord Derby tossed a coin over the naming of a race that was to become the most famous of all races and Lord Derby won! William Frith painted a painting full of atmosphere of Derby Day, which was exhibited at the Royal Academy in 1858 and was so popular that a rail had to be put in to keep back the crowds. Sir Charles Bunbury did manage to have his own race – the Bunbury Cup which is contested on the final day of Newmarket's three-day July Festival meeting and he also introduced two other classics the 1,000 Guineas and the 2,000 Guineas to the Racing Calendar.

The first races were started very simply by a man shouting "Go!" then moved on to a man dropping a red flag. Starting gates were invented in the West in 1939 by a man called Clay Puett and first used in Canada. Another version of these gates was developed for dog racing.

Then to keep a record of all thoroughbreds in the country and incidentally the breeding of race horses Weatherby's was formed in 1770. The results were published in the General Stud Book, which has been updated and re-printed ever since.

By 1800 the five classic races were the St Leger (1776), run at Doncaster, the Epsom Oaks (1779) and the Derby (1780) at Epsom, the 2,000 Guineas for colts (1809) and the 1,000 guineas for fillies (1814) both run at Newmarket.

In 1839 the Grand National – the famous handicap steeplechase was first run at Aintree, near Liverpool. The jumps are huge and growing hedges with names such as Becher's Brook, the Chair and the Canal Turn.

The market for fast racehorses was clearly developing fast, and so too was the relationship that remains synonymous with horseracing today; that between the sport and gambling. At first, matching two horses was the overriding method with which to bet on race horses. One particularly fast animal would be challenged against another and for those that could afford it a wager on such events proved a popular diversion. Betting grew and developed and by the end of the century, the first bookmaker, Ogden, was set up at Newmarket.

Amongst the most celebrated racing horses is Eclipse, the exceptional stallion which never lost a race and about which was coined the famous phrase 'Eclipse first, the rest nowhere!' He won over eighteen important races and retired for stud purposes partly because there was a singular lack of competition. Over 80% of modern racehorses can trace their descent back to him. One famous grandson was Wellington's famous charger Copenhagen. Wellington maintained he never used whip or spur on him, either.

Eclipse was painted several times by the great horse painter George Stubbs. When he died he was attended by a French vet called Charles Vial de St Bel who had escaped the French Revolution. After his death, Eclipse was dissected to try to work out the secret of his success. St Bel weighed Eclipse's heart and found it weighed a massive 14lb, so it was decided that his huge heart pumped blood around the body more effectively, while his back legs gave plenty of leverage. Powerful lungs completed the winning combination. His skeleton is still owned by the Royal College of Veterinary Surgeons and can be seen at the National Horseracing Museum in Newmarket. St Bel was not in England just to dissect a racehorse however famous, but to start a Veterinary College and this he achieved and the College was founded with the help of members of the Odiham Agricultural Society in 1791. It received its Royal Charter in 1844. Then in 1853 it acquired its first home in Red Lion Square. Up to then it had held its meetings in the Freemasons Arms. Since 1995 it has been housed in Horseferry Road.

Frankel is the Eclipse of modern times. Foaled in 2008 and trained by the famous Sir Henry Cecil he won all his 2 year old races and when 3years old was judged to be the best of his generation and the best race horse in over sixty years. He has now been retired by his owner Prince Khalid Abdullah, to stud after being the unbeaten winner of fourteen races. He was confirmed as "The New Benchmark of Equine Excellence."

Racing is very popular in France mainly at the famous race tracks of Longchamp, Saint-Cloud and Deauville. But racing takes place all over the continent. Italy has its own very special race the Palio di Siena which takes place twice a year and is all caught up in pageantry. There is also the Italian Derby held at Roma Ippodrome Capannelle which lies in the hills south of Rome near the Appian Way. Greece, once the mother of horse racing is now considered to be not up to international standards. The decline began even in Roman times.

The first race track in the United States was opened on Long Island in 1665. And in 1868 the Americans set up their own stud book and in 1894 their own Jockey Club.

The heroine of the American racing world was a filly named Ruffian (1972-1975) nicknamed "The Heartbreaker" who ranked among the top US champions of the C20th. She won the Triple Crown and won 11 out of 11 races but tragically when racing against the winner of the Kentucky Derby she snapped the sesamoid bones in a front forelock but refused to stop the race. She won and the vets performed a four hour operation but in vain she struggled to get going and tragically had to be put down.

In the Arab states racing is tremendously fashionable and in Qatar famous races such as "The Emir's Cup" and "The Seniors Single Championship" and "The Sporting Spirit Cup" are run annually. Doha also holds events such "The International Horses Beauty Contest".

In the Far East, racing was introduced into India, Pakistan and Hong Kong by the British. It was banned in Mao's China and the sixty year ban only ended in 2011 with a race course being built in Wuhan, a city on the Yangtze River, with a history stretching back 3,500 years. It started the overthrow of the Qing Dynasty in 1911 and now is one of the most important Chinese cities and centre of transport. Its nickname is "The Chicago of China". The racing is very international and recently they invited seven European lady jockeys from assorted countries to take part in racing.

Mongolia, the origin of riding, holds a very ancient and important annual festival, Naadam, which figures the "Three Games of Man". These are Archery, Wrestling and Horse Racing all taking place before very modern grandstands. The horses are not thoroughbreds but strong Mongolian breeds who race over long distances, which vary according to the age of the horses.

Another Mongolian race started in 2009 is the Mongol Derby which at 1,000km is the longest and toughest horse race. It follows Genghis Khan's postal route of 1224 AD across the Mongolian steppe. In the same way it uses a network of 25 horse stations where the riders pick up the next tough little semi-wild horse for the next lap. The riders spend several nights on the way, camping or with nomads. In 2013 the gruelling race was won by a nineteen year old English woman Lara Prior-Palmer.

Another sport grew from hunting – that of coursing. Greyhounds have always been the dog of choice for coursing and in 1776 a code was written for the sport at Swaffham. But even before that under Queen Elizabeth I the Duke of Norfolk drew up some established rules. In 1836 the famous Waterloo Cup race meeting for coursing with grey hounds was instigated in Lancashire. Like hunting this event too fell a victim to the League against Cruel Sports and sadly held its last meeting in 2005.

However one of the greatest modern sports is that of greyhound racing, a sort of offshoot of coursing. Amazingly the modern form was started in the USA and not England in 1925 by Owen Patrick Smith who also invented the first mechanical lure to run round a circular track. Smith took it over to England and the GRC (Greyhound Racing Club) was established, and the first race meeting was held in a stadium near Manchester in 1926 only a year after the American races had begun.

Another form of racing is with camels, which are trained from the age of two years old. Just like horses, their breeding is carefully monitored and training starts when they are two years old. Also like racing horses they can be sold for extremely high prices. The start of modern camel racing was overseen by Sheikh Zayed, the former president of the United Arab Emirates, who on a scale unknown to European horse racing, personally owns 14,000 camels. There are now 15 racing tracks across the UAE with all amenities laid on and there can be over 50 camels in a race. The top race is the King's Cup, run in Dubai. Islam is, in theory, totally against gambling, but the winners receive magnificent prizes such as luxury cars. Another big difference to Western racing are the jockeys, who until recently were very young boys about 6 or 7 years old, or even younger, chosen or even bought mainly from South Asian countries for their light weight. These child jockeys were often injured or even killed. Child jockeys became an international issue and eventually the use of children under 15 was banned and now jockeys are monitored and licensed. However the latest phase in the jockey scandal is the use of robotic jockeys which by passes all the regulations, weigh hardly anything and are equipped with whips which can be activated using remote control by the owner driving alongside the track in his 4 X 4. This has been going on since 2006. Now masses of cars drive alongside on several roads to view the sport. The speed of camels being especially bred for racing has increased 30%, as compared to horses which has only increased a few per cent, but then horses started as a fast moving animal as opposed to the camel originally plodding across the vast expanses of desert.

An unexpected innovation came through Eclipse – the very first form of horse transport. Eclipse's owner in 1769 was an obese heavy drinking Irishman, Dennis O'Kelly who had neglected Eclipse's feet which were in such poor condition they could not cope with long marches from race course to race course by road. Luckily Dennis's brother, Philip devised the earliest form of horse box for Eclipse to ride in, consisting of a four wheeled carriage with two horses to pull it. This was the first known transporting of a horse overland.

Surprisingly this invention did not immediately catch on and it was forty years before another horse was conveyed by horse box. In 1836 a racehorse called Elis was transported 170 miles from Hampshire to Doncaster to run in the St Leger which he won adding another to his score of eleven wins out of fifteen. But old fashioned trainers complained against the "conveyances" in no measured terms, insisting that such an "unnatural mode" of transporting horses was "decidedly injurious to the ticklish constitution of the trained running horse"

Then for a while, rail was a preferred method of transport not only for horses but for general livestock, who benefitted from avoiding long walks to market. It seemed the cheapest and fastest mode of travel to get race horses from one race course to another. But trains are notoriously noisy forms of transport and soon demonstrated that they were most unsuitable for highly strung horses, and in 1905 J. Wortley Axe, the President of the Royal College of Surgeons wrote that everything in a train seemed designed to make alarming noises and advised against their use. Then gradually horseboxes came more and more into use, and by 1912 Vincent horseboxes were being made in Reading and now many makes of horsebox are a common sight on the roads of England.

Moving horses by sea has always been a risky business especially on long voyages or rough seas as horses, owing to their stomach arrangements, cannot be sick and often die in turbulent conditions on a voyage. Now for moving horses to race or for breeding, the new modern method is air travel. Now the modern fashion is for horses to attend races all over the world.

On the glamour side of our use of animals, horses are the most familiar in processions. They were and are the ultimate power machine for hunting, for racing and for playing dangerous games. Utilitarian needs were of prime importance but display and showing off were not far behind. In fact ceremonial trappings were linked with horses from the very beginning – and still are. In the regal horse graves of Pasryk in the Siberian Altai Mountains the most beautiful saddle cloths, embroidered in silver and gold thread have been found dating from C4[th] BC, and also even more exciting face masks specially for horses, known as *chamfrons*. These masks resembled birds of prey and reindeer (the horses' predecessors in the north) and were meant to make the horses look taller and more terrifying.

Adornments of all kinds were developed for horses in the equestrian world, down the centuries. For example the Moguls of India who were the inheritors of Genghis Khan's Mongols used yak tails, sometimes as many as seven, as decoration on their horses' saddles. For ceremonial occasions and riding in processions, caparisoned in gold and silver and rich fabrics horses were and still are the ultimate statement of awe and beauty. An astonishing variety of ornaments still proudly adorn the horses in the British Royal processions and any great horse event. Horse drawn coaches like the Queen's Coronation and Jubilee carriage are far more impressive than a line of cars would be, and the very word cavalcade is derived from 'cavalry'. The horses are trained to cope with the noise of a great public by simulated noises. The famous draught horses are the Windsor Greys standing at 16 hands which have been stabled at Windsor since Queen Victoria's day. One of the great annual events is Trooping the Colour which takes place at Horse Guards Parade. This started in about 1,700 under Charles II and is based on the necessity of showing the flag to the troops so they could rally to it in the heat of battle. The modern parade consists of about 1,400 men, 200 horses and 400 musicians from 10 bands. The march starts at Buckingham Palace processes down the Mall to Horse Guards Parade and back again. The Queen took the salute riding side saddle until 1986 when her favourite black mare Burmese died.

Funerals also provide an excuse for pageantry for anyone willing to pay. One of the most striking sights is a caparisoned horse being led with the dead owner's boots reversed in the stirrups. This custom is supposed to date back to Genghis Khan or even earlier and represents the rider looking back at his troops for the last time. A relatively modern example was Old Bob, Abraham Lincoln's horse which took part in this way at his funeral in 1865.

Another tradition is that on election, the new Pope would lead an elaborate procession either riding a white mule or being carried in a litter from the Vatican to San Giovanni in Lateran, which is the ancient site of the Papal see. Pictures of these processions showed the crowds zigzagging in what was known as the bustrophedic pattern which is a term applied to languages written with the alternate lines from left to right and back again in the same way as a field is ploughed. The Popes in the C16[th] celebrated the Feast of the Annunciation or Lady Day leading a procession through the streets of Rome, also usually riding a white horse or mule. But the 'Eminentissimi' generally were not good horsemen and so were frequently tied onto their saddles.

Horses really do seem to play up to the public. Probably the most exciting representation is that is of the Panathenaic procession on the Parthenon.

> *"The horses that guide the golden eye of heaven*
> *And blow the morning from their nostrils."*
> Christopher Marlowe

Camel fairs are held in several parts of India every year and are great tourism attractions. In Rajasthan a very big camel festival is held at Bikaner every January. The camels are adorned in the most amazing decorations, tassels, pom-poms of every colour and bits of mirror. But very original forms of adornment are patterns shaved in the fur of the camel, not just abstract designs but people, animals and birds. A procession takes place against the red sandstone backdrop of the Junagarh Fort. There are competitions for best breeds and camel dancing demonstrating deft footwork.

Another fair is held in Pushkar every October or November. It is for cattle as well as camels and is combined with a religious festival. Races are held for camels and horses.

The Marwar Festival is the second biggest in India, lasts for eight days and is held in Jodhpur in September/October in memory of the heroes of Rajasthan. It is a real working fest and approximately 70,000 bullocks, horses and camels are traded during it; but it has a lighter side with a camel tattoo and games of polo taking place before the backdrop of the Umaid Bhawan Palace.

Jaisalmeer Desert Festival is held at the end of February every year and lasts three days. The backdrop is the fabulous C12th fortress built of golden sandstone. Apart from folk dancing and singing and the competition for the longest moustache they have camel races, camel polo and even camel dances.

Another animal used in pageantry and ceremonies is the elephant, often forming an integral part of royal procession, frequently carrying howdahs on their backs. They seem to be naturals when it comes to decorations. Their size and colour lend themselves as a canvas for fantastic painting. Their long trunk is perfect for decorating with gold-plated caparisons. Bells, yak tails, peacock feather fans, silks and the mahouts in costume wielding colourful umbrellas are all used to adorn the elephants. Families specialize in designing and making all the ornaments needed for the numerous festivals. And to this day they take part in religious processions and games. Kerala in South India is the scene of several of these including one featuring amazing coloured umbrellas. A relatively new festival is held at Surin in Thailand, during which there is an elephant round up and an Elephant Breakfast.

Not all the mahouts are kind to their elephants which are sometimes rented out for the occasion for large sums of money. The elephants often have to endure heat, noise and sometimes having to go without food, drink or sleep all day. Not surprisingly the elephants sometimes retaliate with fatal results.

One of the main festivals is Diwali which is celebrated all over India and part of it is the worship of Lord Ganesh, the elephant god of prosperity and wisdom, who, it is believed, descends to earth to bless his followers.

Other uses for elephants? Circuses were a popular one in Roman times, as they are or were in modern times. But there was a vast difference…then the elephants were made to fight each other in gladiatorial combats. They were given stimulants to fuel their anger, but when lions and tigers were unleashed upon the elephants, the crowds roared their disapproval. The elephants had won their hearts. They no longer figure in today's circuses, as much cruelty has been uncovered and animal rights are gradually halting this practice. So really it is now mainly in India that you see the elephant dressed up at his most glamorous for the numerous festivities.

Probably the best known elephant festival takes place in Jaipur annually on Holi eve. The elephants are highly decorated and painted and wear wonderful embroidered and tasselled hangings. They process up and down and parade before an admiring audience. However the elephant part of the festival was cancelled in 2013 owing to pressure from animal rights.

Also in March there is the Parapally Gajamela or elephant pageant at Parapally with fifty elephants caparisoned in golden headdresses that stretch down their foreheads and top of their trunks and process through the crowded streets. Kerala holds a similar festival every year in mid-January

Cattle have their cowbells, which were originally to show where they had wandered off to. The earliest evidence of them comes from pottery bells from 3rd millennium BC in China. But they are very much used in the Alps for transhumance. In spring when the snow has melted sufficiently for the cows to go up to the Alpine meadows and high pastures an event takes place in most Alpine villages. The cows are led in a procession called Alpaufzug tinkling their bells and with wreaths of flowers decorating their horns. Often the best milk cow leads the procession with the biggest bell. In autumn, when the cows come down again similar events take place called Alpabzug.

The ancient sport of Bull fighting has a long and varied history. Fighting between bulls was depicted on the walls of Egyptian tombs from the time of Narmer's son Aha (3,000BC) in the 1st Dynasty. A form of Bull-leaping is still practiced in various towns in Gascony, in SW France as part of local summer festivals. It is known as the 'Course Landaise' and young cows are used instead of bulls. A team of leapers "sauteurs" and dodgers "écarteurs" are all dressed up in fancy embroidered waistcoats and evade the charging cow with a wide variety of acrobatic dodges. The cow is restrained by a rope tied to her horns and is not to be injured, but of course accidents do befall the leapers. There is a more dangerous form in Spain known as the Recortadores, where bulls are used instead of cows and the athletes as well as dodging and leaping, vault over the bull using a long pole. Another difference is that the bull is unrestrained by a rope, making it all a much more dangerous sport and more akin to the real Spanish Bull Fight.

Spain's famous sport/spectacle the Bull Fight originated in 711 with the first fight taking place at the Coronation of King Alfonso VIII. It is believed to have its origins in early hunting practices but of course it has developed a form of its own with Picadors on horseback making the first encounters with the bull, followed by the Matador on foot playing the bull with his cloak. The modern form originated in Ronda, a town in Malaga and the rules were laid down by Francisco Romero c.1726. The town possesses the oldest bullring dated 1785. The breed of bulls used for the fighting is very ancient, called *Toro bravo* and descended from an ancestor *Bos urus*, possibly a variation of the aurochs and pretty well untameable. There were once 600 bullrings in Spain. Cesare Borgia, the son of the Pope Alexander VI, was an eager bull-fighter and even cut off a bull's head in the course of one encounter.

Bull fighting is still very popular in parts of Spain. Animal rights activists are protesting about the tradition but another way of looking at it is that the bulls have several years living in enviable conditions before a short sharp fighting end when high on adrenaline, probably less stressful than a long journey in crowded cattle trucks and trains to the slaughter house. If the sport were totally banned, then this breed of *Toro bravo* would become history. Catalonia voted to ban bull fighting in July 2010 and this ban came into effect in January 2012. Matadors themselves describe bullfighting as a dramatic ballet dance with death itself.

Oman has a very different type of bull fight in which men are not involved as it is a fight between two Brahmin bulls. The fights take place in the fort of Barka and the loser is the one which gets knocked down or runs away. This role for bulls supplants the original use for them, pulling ploughs and turning water wheels.

Another form of bull fighting is Jallikattu played mainly in the Indian state of Tamil Nadu, which is quite different to the Spanish form. The bulls are chased not killed, and in earlier times it was a way of women choosing their husbands. Rock paintings from 3,500 years ago show men chasing bulls. Like the Spanish bulls, the Jallikattu bulls are raised in almost wild conditions. As with Spanish bullfighting, the question of cruelty to animals has been raised. And also the same end would come if the sport were to be banned – the extinction of an ancient breed of cattle.

In the American rodeos, cowboys have to ride bucking broncos or sometimes bucking bulls, from which they lasso calves or steers. They have to drive wagons and perform all the tasks that cowboys or vaqueros have traditionally done. Probably the most famous is the Calgary Stampede which is held in Canada and brands itself "The Greatest Outdoors Show on Earth." The stampede takes place annually in July and lasts ten days and is attended by over a million visitors. Some of the main attractions are bull riding and steer wrestling. Prizes of $100,000 are awarded for these dangerous sports but they are now also being threatened by animal rights, and possibly 'Health and Safety'.

Not exactly a game but a great spectacle – the Spanish Riding School was founded by Archduke Charles in 1580, and the Hapsburgs bred the Lippizaners, the famous white horses from just six stallions. They are trained with Xenophon's recommended tender care for their exacting performances, which are to a degree, based on fighting horses but are mainly to strengthen the horses' muscles and teach it coordination and obedience. François Robichon de la Guérinière (1688-1751) had a similar approach in one of the world's greatest books on horsemanship 'Ecole de Cavalerie and his views are the basis of the training in the Spanish School. The final step in their training 'The Airs above the Ground' takes two to four years training and is performed to classical music. Performances can be seen at the Winter Riding School at the C18th Hofburg Palace, Vienna, in an enormous room just like a ballroom except that the floor is covered in soil and the viewing public gazes down from balconies all round.

Manali is an Indian hill resort where they ski in winter. Yaks are used as an imaginative ski lift in the following way – the skier is standing at the bottom of the slope and the Yak at the top and they are connected by a rope that runs through a pulley. The unusual apparatus is activated by the skier shaking a bucket of pony nuts which he puts down hastily as the Yak takes off down the hill for his snack and the skier is dragged up the slope. It is difficult to see how it continues from there.

---ooO0Ooo---

ETCETERA

1. The Queen's horse Estimate won Royal Ascot's Gold Cup in 2013. It was the first time a reigning monarch has won in the race's 207 year history and the Queen was totally overjoyed.

2. Aristophanes, the comic playwright, describes the troubles of a father whose son has too-expensive tastes in horses: "Creditors are eating me up alive...and all because of this horse-plague!" (Aristophanes, *Clouds* l.240ff.)

3. One special incident of a dog in a parade was that of Caesar the wire haired fox terrier beloved of Edward VII who walked in his funeral procession ahead of nine kings and more than a dozen heads of state.

---oooO0Oooo---

CHAPTER **12**

WAR

Knight in Armour   Boer War Memorial, Port Elizabeth

"Cry 'Havoc' and let slip the dogs of war"
Shakespeare. Julius Caesar (Act II Sc III)

It is forbidden to kill; therefore all murderers are punished unless
they kill in large numbers and to the sound of trumpets.
Voltaire

Ploughing was essential, draught and riding were crucial, games and racing were fun, but Man's passion for fighting dragged all these six animals out of peaceful employment into wars in which they were to be used in all manner of ways. There is no archaeological proof but no doubt prehistoric fighting over kills took place with man-wolf teams fighting each other for the meat and these fights began to escalate. Animals in warfare are such a vast subject that we can only select a few examples down the millennia.

226

Some of the earliest employment of dogs was the use of them by armies on the march. There is only so much living off the countryside to be had and so the army took flocks of sheep and herds of cattle with it – a giant larder on the hoof. And who helped drive the animals and guard them from predators and robbers? Why dogs, of course, one of the breeds was actually an early form of Rottweiler. But dogs did not just take a passive role in warfare. The giant Molossian dogs were bred not just for use in hunting; they were also employed for actual warfare by the Persians, the Egyptians, the Greeks and the Hyksos in Egypt. They were protected by chain mail and a spiked dog collar and would take an active, aggressive part in the fighting. Hammurabi used them as far back as 2,100BC. Another mention of dogs in war was the use of a canine battalion in 628BC by the Lydians, who lived in what is now Turkey. They were trained to fight also wearing spiked dog collars and even armour. Cambyses II, son of Cyrus the Great used them against Egypt in 525BC in a rather cunning way. He used the Egyptian reverence for animals against them by having the images of their most revered animal the cat painted on their shields and driving dogs and sheep before the advancing Greek army.

When the Romans conquered Britain (43AD) they came across Pugnaces Britanniae, possible the ancestor of the mastiff, huge with wide mouths. They were used in dog fights in the Roman amphitheatre and the Molossus dogs were no match for them, according to the author, Gratius Falsius, writing in 8AD. If the Pugnaces had been running the war the Romans would never have succeeded in their conquest. The British dogs were reported by various writers including Strabo, the Greek historian and Tacitus the Roman writer.

In the 1500s the Spanish conquistadores used fighting dogs, mostly mastiffs, wolf hounds and sheep dogs and together with horses, they terrified the Aztecs and Mayans who had never seen horses before nor domesticated dogs, trained in warfare. Another fighting dog was the Irish Wolf-hound, which had been trained to drag the invading Norman armoured knights off their horses and if they weren't killed by the heavy fall then the Wolfhound, sometimes in pairs would finish them off. In Tudor times, both Henry VIII and Elizabeth used fighting dogs, the latter sending several hundred of them to help quell the two Desmond

Rebellions in Ireland at the end of the C16[th]. In 1525 Henry sent 400 mastiffs to support Spain.

In the First World War thousands of dogs were used in various capacities from giving warning of the enemy approaching to carrying messages and taking medical kits to the wounded; just as they had been used for that purpose by Frederick the Great in the Seven Years War and by Napoleon on his campaigns. A descendant of the Irish Wolf-hound is the Bouvier des Flandes, which was a farm dog, a cattle dog and often pulled farm carts but was then trained to haul heavy machine guns to the battle fronts Apparently up to 20,000 dogs were taken from dog pounds, donated by their owners or recruited from the police force and trained to carry out these duties. But not so obvious must have been the comfort and companionship they gave the troops in their grisly trenches.

In World War II the Russians trained dogs to seek food under tanks, having strapped explosives to their backs. Driven by carefully induced hunger a dog would pass under the German tank and the tilt fuse would be activated by the tank moving over head and set off the explosive which the unsuspecting canine suicide bomber was carrying. As they couldn't differentiate between one tank and another they sometimes blew up their own Russian tanks.

In the Vietnam War the Americans employed over 10,000 dogs, of which only a few hundred made it back home; and the Russians used them in the Afghan War in the 1980's. Their main jobs were scouting and smelling out booby traps, mines, snipers, caches of weapons and fugitives. Amongst their duties were laying telephone cables, rescuing the wounded from the battlefield and sentry duty. They also acted as mascots and helped build morale in the troops. In the Vietnam War about 5,000 dogs were used and the K9 units are estimated to have saved about 5,000 lives. In the Iraq war the USA used them to intimidate prisoners at Abu Ghraib and Guantanamo Bay and that was only too easy as most of them were Muslim, with their inbuilt horror of dogs.

Dogs, mainly Spaniels and Labradors, proved to be invaluable in the present conflict in Afghanistan. They are expert at sniffing out IEDs (Improvised Explosive Devices) that were causing so many casualties, and just about the only part of the army that is expanding is the RAVC (Royal Army Veterinary Corps). The dogs are trained to track down a variety of explosives for which service they are rewarded with a ball game or food. They also have to learn to withstand the terrible noises of war – crashes, explosions, helicopters. One was flown into Osama bin Laden's hideout with the US special forces and was strapped to a US Navy Seal as they abseiled out of the helicopters. They were also reportedly used to capture Saddam Hussein in 2003. They can be fitted with cameras to alert their handlers of possible dangers.

As a not unexpected by-product, dogs also do a tremendous job in raising troops' morale so not surprisingly every field commander is very keen to have dogs and their handlers. Many dogs have been awarded the Dickin Medal which is the canine VC, instituted by Maria Dickin 1870-1951, the founder of the PDSA, in 1943 and so far has been awarded to eighteen dogs and three horses and a number of pigeons and even a cat. The latest receiver of the Medal, posthumously with her handler was Sasha, a yellow Labrador. She had been sniffing IEDs in Afghanistan but they were both shot by the Taliban in May 2014. Training for the war in Afghanistan took place in Norfolk where a mock-up Afghan village, complete with mosque, bazaars and smells was built in 2009 at the cost of £14 million for use of the army to prepare troops as well as dogs.

Our next animal to be domesticated was, of course, cattle but they do not seem to have played an active part in warfare. Of course they were used as draught animals for supplies and for food on the hoof but there are no glamorous stories of men galloping into battle astride ferocious gaurs or aurochs, probably because they would have been literally riding a two edged weapon. But a warlord was frequently called "Lord of Cattle" and many raids were made in search of cattle and these raids were blessed and justified by the priests.

Then the Horse! Once Man and Horse combined War rose to dizzying heights and descended to terrifying depths for over 5,000 years. The Horse as Warrior was *the* Horse for thousands of years. All of Europe and a large part of Asia became obsessed with the Horse as a weapon. This was an amazing feat centred on man's ability to train the Horse to go into battle when his natural defence and reaction to danger is flight – to escape with the speed Nature endowed it. The funny little Przwalski type ponies were transformed by breeding and training into the essential part of the war machine which began as Horse and Chariot and developed into cavalry.

The newly invented chariot was seen as a symbol of power, a symbol of speed and it revolutionised war for over four thousand years. Gaining the psychological advantage is usually the deciding factor in war and the fatal combination of Horses and Chariots were certainly being used by 3,500BC by the Persians. This arrangement had replaced the ridden animal to some extent and had become real instruments of terror. It seems strange that driving was preferred to riding for so long but by 2,000BC chariots rolled from Egypt right through Greece to India. The Aryan word for 'chariot' is 'ratha' which is the root of words roll and rotate. Possession of a Horse became a status symbol a "must have" for the nobility. Class distinction was easy to assess – there were the equestrians and then there were the pedestrians. Examples were the tribes in Kazakhstan where those who had battle chariots defined the deep social divide between the ordinary people and the chariot warriors. Not surprisingly the latter's guardian was the Sun God, so often connected with horses who had pulled the Chariot of the Sun in mythology. The Greeks called him Pegasus and described him as:

> *"My horse a thing of wings, myself a god."*
> *Wilfrid Scawen Blunt 1840 – 1922*

> *"The Horse is a warrior and a foreigner."* said the
> soothsayer to King Croesus of Lydia.
> *"The armoured Persian horsemen and their death-dealing*
> *chariots were invincible. No man dare face them."*
> *Herodotus (484-430BC)*

Again and again, in Egypt, Greece, China, India and Rome, the barbarian is depicted being crushed under the wheels of chariots. But the real advantage of the chariot was that it acted as a moving platform for archers, who were protected up to the waist by the semi-circular shield protector on the front of the chariot. The Chinese designed the best version of the chariot and had a standing army of 10,000 of them.

One of the most fearsome innovations was the scythed chariot which carried long blades horizontally from the axles. These were introduced by the Persians in the mid C5[th] BC. The chariot had a driver and carried two warriors, which must have been quite tricky to operate. There was a custom, which has inadvertently helped archaeologists, to bury the warriors with their chariots and even with their horses

Then after all those thousands of years the chariots were ousted by another huge transformation – from chariots to cavalry. This was another innovation of enormous importance. According to Herodotus as early as the C5thBC, horses were just beginning to be bred larger and stronger in order to carry heavier men and their armour and so chariots were being phased out in favour of cavalry although they did linger on in such far-flung outposts of the Roman Empire as Britain, until C1st BC.

This combination of the Horse together with their genius for archery led to a new and terrifying phenomenon arising – the Scythians. They were a nomadic people who moved into Russia from Central Asia in C9th – C7thBC. This was a collective name for the men of central Asia, whose new mastery of the art of riding horseback gave them dominance over all they came upon. No wonder these unstoppable fearless riders on their speedy and elegant mounts gave rise to the myth of the centaurs. Incidentally they were the first people to wear trousers, so different from the tunics or flowing robes of the time, and so much more suitable for riding astride. (The only others who took to trousers were the Ancient Chinese cavalry, introduced by King Wu of Zhao in 375BC. and the Achaemenid Persians) Also to the astonishment of the Greeks, the Scythians had another innovation; they castrated their Stallions to make them easier to control. They were freemen and received no pay other than food and clothing. However on production of a head of a slain enemy they could share in the distribution of captured booty. However it wasn't all roses. Many male Scythians became impotent from the constant jolting of their testes on the horse. And Herodotus mentions a 'female illness', obviously bleeding of some sort that could have been caused by piles or fistulae. (These troubles also appeared with knights during the Hundred Years War (1337 – 1453) being called 'androgynes'.)

So by the end of the C6th BC the Scythian superior horsepower and fighting ability had enabled them to sack Nineveh, the seemingly unassailable Assyrian capital (612BC). They then went on to defeat Cyrus the Great of Persia, whom they killed in 530BC. As a token of their savagery, Tomiris, the Scythian Queen gave the bizarre order that his head be put in a wineskin filled with human blood.

In the Old Testament Jeremiah, no less, gave dire warnings about the approach of the Scythian hordes. He spoke of whole cities in Palestine who would "flee for the noise of the horsemen and the bowmen, a mighty nation, whose language thou knowest not. Behold a people cometh from the north country."

War figures largely in other books in the Old Testament of the bible and a good selection of equestrian quotes comes from Job 39:

19. *Hast thou given the horse strength? Hast thou clothed his neck with thunder?*

21. *He paweth in the valley, and rejoiceth in his strength: he goeth on to meet the armed men.*

22. *He mocketh at fear, and is not affrighted; neither turneth he back from the sword.*

23. *The quiver rattleth against him, the glittering spear and the shield.*

24. *He swalloweth the ground with fierceness and rage: neither believeth he that it is the sound of the trumpet.*

25. *He saith among the trumpets, Ha, ha; and he smelleth the battle afar off, the thunder of the captains, and the shouting.*

Warfare had now become a truly nomadic activity. It culminated in the Huns, the Phrygians and the Mongol hordes under Genghiz Khan. It seems that only settled people can make civilisation – and then only sometimes.

It is difficult to visualize the terror the hordes inspired. Imagine the ordinary people on the steppes, going about their business of farming when suddenly over the horizon, in a cloud of dust comes a multitude of galloping horsemen screaming war cries to pump up their adrenaline. Perhaps it can be compared to the great tank squadrons of the Germans and the Russians in the last war unselectively and impersonally annihilating all before them.

Ancient Greece only had infantry to begin with, but urged on by Xenophon the Athenian writer, who was extremely interested in horses and horse training, soldier units of cavalry began to be trained.

Philip of Macedon and his even more famous son Alexander the Great (356-323BC) used mounted soldiers extensively in a cavalry divided between the heavy *hetaroi* who carried heavy thrusting spears and the lighter *podromoi* who were used more as scouts and were lightly armed with javelins. Alexander's first encounter with the Persians was the Battle of Granicus (334BC) when he landed with a large army and 5,000 Companion cavalry, consisting of eight squadrons led by Philotas. This battle was the opener in Alexander's bid to liberate the Greek cities of Asia Minor from Persian domination. The most famous was fought in 331BC – the Battle of Gaugamela. Darius had prepared the terrain for his scythed chariots, camels, cavalry and elephants, but Alexander carried the day and Darius and much of his cavalry fled the field of battle. Alexander was crowned "King of Asia".

Alexander was obviously a great animal lover. He managed to tame Bucephalus which means Oxhead, a horse brought to his father Philip of Macedon and seemingly untameable. The story goes that the twelve year old Alexander noticed the horse was shying at his own shadow so he led him into the sunshine and talking quietly to him, managed to mount and ride him. When his horse was killed after many battles at Hydaspes, Alexander was heartbroken and named a city after him – Bucephela which is now Jhelum in Pakistan. He also owned a dog called Peritas who went through numerous campaigns with him and there are several stories about how the dog saved his life; one claims that the dog bit the lip of an elephant that was charging him and another that the dog went to Alexander's aid when he was cornered by the enemy Mallians and killed several before being fatally wounded himself and dying with his head on Alexander's lap.

In 330BC Alexander the Great plus 60,000 troops sacked the Persian capital of Persepolis and used 10,000 other baggage animals and 5,000 camels to carry away the spoils.

The early cavalry of the Roman Republic were more status symbols of the social elite than of practical use, so mercenaries from various conquered territories were employed to do the actual fighting as opposed to the ceremonial strutting of the Roman *equites*. But gradually the cavalry began to fulfil a more important role in the Roman Army, which was mainly dependant on its legions. The heavily armoured cavalry were called Cataphracts and each rider carried a lance called a '*kontos*' and sometimes a sword called a '*spatha*'. The armour was scale armour and looked not unlike the scales on a reptile. All this was in answer to the brilliant Parthian horsemen who darted in and out completely overwhelming the Roman infantry standing in immobile close order. The Roman cavalry had become a striking force and was not just working behind the scenes carrying messages, scouting and performing the odd raid.

The cavalry wielded by the Persians and Byzantines, similar to the Cataphracts, were the Clibanarii, which was a sort of nickname meaning 'camp oven' owing to the appearance and effect of their armour. It was under the Emperor Hadrian that the cavalry or Cataphracts really caught on and developed through the 3rd and 4th centuries in the endless fighting to repel the barbarians. The measure of a first-class Cataphract was if he could impale two soldiers on his one spear during a charge.

When the Romans arrived in Britain in 55BC with Julius Caesar, the Celts were riding small ponies which were hopeless in war, so it was a walkover. It was the same again when Claudius arrived with his ceremonial elephants. For the famous revolt by Boudica and the Iceni, she was probably driving a tiny chariot drawn by these ponies. It was not until the time of Alfred the Great that studs for better stock were introduced into Britain.

It was the Hun archers who riding into Rome not just on horse-back but using stirrups and saddles which led to Rome's downfall. The Huns used these stirrups and saddles as a standing platform from where they shot arrows. Rome's last battle was against Attila the Hun in 451 AD; by 470 AD the Roman Empire was finished.

Then a new force came from the east. The new religion and leadership of the prophet Muhammad (570AD – 632AD) changed his followers from herdsmen into equestrian warriors by promising them paradise as reward for being slain in battle – a forerunner of the suicide bombers of today. He declared that the horse's back was the human seat of honour but he didn't do as much to advance the actual breeding of better horses as Solomon. His own personal horse was called Al Borak, Arabic for 'lightening'. When he died in 632AD his followers said he was carried to the seventh heaven on Al Borak's back. From Mohammed there are more then 2,000 mentions of the horse, which of course conferred on the animal an almost sacred standing. One lovely quotation from the Koran...*"Horse, you are truly a creature without equal, for you fly without wings and conquer without sword."*

One of the most critical battles was the Battle of Tours (or Poitiers) when Charles Martel – the Hammer defeated the invading Islamic army of Abd Rahman Al Ghafiqi in 732AD. This was the battle that saved Europe, for the second time, from the relentless spread of Islam. The Franks were infantry only and in spite of that defeated the Islamic cavalry. Martel was so impressed by their performance that he was really responsible for introducing cavalry into Europe. He considered the use of the stirrup so important that those riders who brought to his army a horse, saddle and stirrups were rewarded with a title to church land, which he had stolen. He realised that the new invention was pivotal for keeping the heavily armoured knights from toppling off their horses. He could be said to have introduced feudalism into Europe.

Riding a horse into battle was the ride for the gods and the Kings and the aristocracy. And as Shakespeare's Richard III cried out in desperation at the Battle of Bosworth, having been unseated: *"A Horse! A Horse! My Kingdom for a Horse!"*

"Cavalry" is an evocative word, a word that conjures up thousands of battles all over Eurasia going back thousands of years. The word is derived from the French 'cavalerie', quite modern but it refers to the very early times of man and horse-drawn chariot and man riding horse into battle. Against such combinations the foot soldier was at a great disadvantage.

There were two types of cavalry the Light and the Heavy and they developed along very different lines and in different parts of the world. The light horsemen of the East were essentially horse-archers. They had smaller faster horses and galloping from left to right (to suit their bows) along the enemy letting fly numerous arrows and ending with the notorious Parthian shot over their horses' tails as they galloped away.

Taking a horse into battle, you had to choose between speed and the ability to carry heavy weights such as a knight in armour. The average horse can carry 25% of its own weight, so one solution was to ride a faster lighter horse to the battle then change into your armour and ride the heavier horse into battle. The Rolls Royce of warhorses was, as mentioned in Chapter 3, the Destrier, also known as the Great Horse.

As stated, a horse's natural reaction to danger is flight, so to be trained for warfare they had to be taught to cope with noises, smells and the clangour of arms. Encountering unknown animals like camels and Elephants would send horses into a panic, so the Persians for example, had dummies of elephants made, smeared with unguents to familiarize their horses to them and overcome their fears. Caesar and Pompey went further and introduced their horses to the real thing. In the war between the Lydians and the Persians according to Herodotus, Cyrus used the smell and sight of camels to rout the horses of the Lydian army.

When the Normans arrived in England in 1066, the Bayeux tapestry shows the Normans on the horses they had brought over by boat and the English fighting on foot, so the Normans had full advantage, and the Battle of Hastings was a ride over.

The mediaeval knight is an iconic figure...a legend...a fairy tale. But the knight in shining armour is nothing without his horse. His rank had become closely linked to chivalry and a special code of conduct and good manners. But when he wasn't singing "chansons" to his lady love or jousting with the other knights, he actually went to war. From C12th to the C15th was their period. Being a knight had its drawbacks as you incurred big expenses maintaining your fighting equipment armour, weapons, and most of all your horses or destriers, known as the Great Horse, and then you had the obligatory forty days military service to your liege lord. Apart from war there were pageants, ceremonies and jousting and the equipment, armour and clothes to pay for.

In the C15th King Mattias Corvinus raised the Hungarian Hussars. The glamorous word 'Hazar' merely means 'twentieth' and refers to the then current law that one man in twenty was required to serve in the army. The Hussars, like their predecessors the Mongols, were brilliant and speedy riders and almost impossible to resist. The French dubbed these lightening attacks "coup d'huzzard".

The heavy cavalry was not interested in speed. Armour was worn and the lance was the chosen weapon to use charging against the infantry. A fierce curb was necessary for control as a bit; a high canticle at the back of the saddle helped prevent riders being swept off. Horses too wore armour and elaborately curved and spiked masks called chamfrons protected their heads. Throughout the Middle Ages the armour got heavier 18 – 32 Kilos as field armour and the very heaviest was 41 Kilos. Cuir bouilli (hardened leather) was very popular. But sometimes if the wearer fell off his horse he couldn't get back on his feet without help and so was often butchered, as he lay helpless on the ground. To keep up with this increasing weight the horses had to be bred stronger and stronger and so in a slow-moving arms race, also from the twelfth to the fifteenth centuries, the Great Horse came into being.

But then they too were outclassed by that amazing Welsh invention – the Longbow, at Crécy (1346). It spelled the end of classic chivalry. Another great battle was Poitiers 1356. Led by the Black Prince, again the English were heavily outnumbered by 5 to 1 and their horses had not recovered from the sea crossing, but it proved another longbow victory. The French horse usage was mismanaged and the French nobles were too snobbish to dismount and fight hand to hand next to the lowly infantry. The third great battle between the French and English was of course Agincourt (1415), which started very unevenly balanced as the English force had been reduced by a combination of disease, food shortage and exhaustion from the forced march from Calais. It began primarily as a cavalry engagement with the French dressed in heavy and cumbersome armour being hoisted onto their horses by cranes. Just how outnumbered were the English is forever being argued but it was at least 4 to 1. The English again used the Longbow to amazing effect and were far more mobile. When a Frenchman fell off, he was all but helpless and a yeoman would open his helmet and cut his throat unless he was worth saving for a ransom. Meanwhile the French horses panicked and were caught by sharpened stakes the English had planted for them. The French were routed with over 7,000 killed and at least 1,500 taken prisoner and the English lost about 500.

Horses were transferred all over the world for the purposes of conquest. Sailing across the often rough Atlantic took a heavy toll of the Spanish conquistadores' horses. They slung them and cross tied them but still many died. But the effect of the survivors on the Aztecs and Amerindians was colossal, as they had never seen horses before and at first they called them Big Dogs. They must have had a similar petrifying effect rather like the Scythians in Central Asia, for the conquistadores rode onto an easy victory.

Four hundred years after Agincourt, another Emperor – Napoleon, agonized by piles, rode his white charger, the stallion Marengo to the battle of Waterloo. Actually Napoleon in an early appraisal of PR or because they didn't mean anything to him, obviously had more than one horse, but he called them all Marengo and the skeleton of one ended up in the National Army Museum in Royal Hospital Road, London. Wellington, with a less painful seat rode his favourite steed Copenhagen. The deciding factors of this battle that changed European history were two men, two horses and the better-bred horseflesh used by the English against the French. One reason the horseflesh was better on the English and German side was that the French Revolution had decimated the good breeding stock in France. On the English side was Wellington's ally Marshal Blücher, whose nickname was famously Marshall Vorwärts (Marshall Forwards) referring to his approach to tactics.

Transporting horses by sea continued to be primitive and cruel. The cavalry horses sent to the Crimea in 1854-6 suffered the most awful conditions – dark and airless holds where they were crudely tied up, bad feeding and care. Disembarkation was equally crude and sometimes horses were literally chucked overboard to swim their way to land or drown. Even the callous and unfeeling Lord Cardigan was shocked.

One of the most interesting studies in the Boer War was to see how the various breeds of horses stood up to the work they were given. The army were mounted on animals drawn from all ends of the earth; great round-hipped English chargers, light wiry Australians, mongrel Argentines, wonderful little Burmese ponies and last, but not least, the Cape horses. A military writer had committed himself to the opinion that the Cape horse was useless for military purposes; nevertheless it was on Cape horses that the Boers did all their work, and the rapidity of their movements showed that they were not mounted on inadequate horses. Their informal method of fighting was far removed from the old European squaring up in formation. It was the beginning of the commandos which culminated in the Royal Marines and SAS who use innovative ways of defeating the enemy, far removed from the old square formation.

When war broke out in 1914 the British Army apparently had only 80 motor vehicles so they were almost entirely dependant on horses. Forage for them was hard to come by and they were often seen gnawing at the wooden wagon wheels. The slaughter of horses in those awful four years was appalling, running to over 8 million horses, 1 million of them from England. The combination of modern artillery and barbed wire were disastrous for cavalry who had long outlived their place in battle. But still, against any common sense, they were used. General Douglas Haig, probably not the brightest of generals actually predicted that horses would continue to be used in warfare. He was right, up to a point, as even in the highly mechanised WWII the Poles used cavalry against the German tanks. After being engaged in the war for less than a month 11 brigades of Polish cavalry had been involved in no less than sixteen charges against the Germans.

Another essential on the battle scene was the offspring of the horse – the Mule. It has terrific advantage over other pack animals, according to figures brought out by the Animal Transportation Officer in 1988. It can carry roughly one-third its body weight and that could be as much as 200lbs – 270lbs. It is less temperamental than a horse and from the advantages of 'hybrid vigour' is stronger all round. Its skin is harder and tougher and so less inclined to chafe, its hooves are much tougher, so it can stand extremes of temperature better and needs less water. Buffalo, elephant, yaks and camels could carry more but not go nearly as far. The Hittites certainly thought so. And even Alexander the Great had a chariot drawn by twelve mules. Another of their strong points is their surefootedness over the difficult terrains of such disparate places as the Himalayas and the African Veld. Their pace is slow but steady and as they require less sleep they can go further in 24 hours. One typical perversity is that they go up hill faster than down. They were used in most of the Wars of modern times, the Napoleonic, South African, Indian and the American Civil War. In the Boer War alone 67,000 mules and 76,000 horses had been bought from the USA. In no previous war had so many horses and mules been killed – an estimated 350,000 horses and 5,000 mules. What with poor management, overloading, and not allowing recent arrivals time to acclimatize the average life expectancy of a horse on the British side was only six weeks. The Boers, on the other hand took far better care of their horses mainly because they knew how and partly because they were difficult to replace. In the siege conditions such as Ladysmith soldiers slaughtered them for meat and even invented Chevril which was a jelly like paste made from boiling the horse and used like Bovril. There is a memorial in Port Elizabeth to the horses that died in the war.

In the Crimea, thanks to Florence Nightingale, mules took on an added duty. They were used to evacuate the wounded, which were carried seated in pairs of panniers called *cacolets*. These substitute ambulances could go where more regular ones could not. Another very famous military use of mules was on the Northwest frontier during the Great Game – that armed rivalry between England and Russia for possession of India and Afghanistan. Here the mules were used to transport the 'screw guns' that Kipling wrote about in 'Kim' and other stories.

The next animal – camels made their first appearance in war rather later than the horse. In their own desert environment they definitely have several advantages over the horse because they can carry a much greater weight and require much less food and water and obviously can deal far better with desert conditions. But away from the desert the horse's speed and greater manoeuvrability gave them the edge, because camels, never having had natural enemies in their environment have never evolved a need to move very fast. Several other things in the favour of camels is that being tall they could be used by mounted archers as a platform to fire from over the heads of the rest of the army and when later the same thing applied to firing guns, it turned out that they were not frightened by gunfire. But their secret weapon, their unexpected asset was that the sight and smell of camels can panic horses when they are not accustomed to them.

The first person to exploit this extremely positive feature was Cyrus the Great of Persia at the Battle of Thymbra against Croesus of Lydia in 547BC. In this important battle Cyrus' cavalry was outnumbered by six to one but luckily one of his generals had noted the terror effect of the camels on the enemy forces. Cyrus immediately ordered the transforming of his baggage camels into what was to be the first Camel Corps in history and the Lydian horses turned and fled. The camel stench had won the day. This victory led to the spread of camels into Asia Minor. Herodotus in Book 1 mentions the same positive use of camels in the Persian army the following year against Croesus at the siege of Sardis when again they were used to rout his cavalry.

Then once the North Arabian saddle had been invented, actual fighting from a camel's back was revolutionized. Equipment could be hung from the saddle and because the rider was now so much more secure, he could use spears and long swords, as well as missile weapons. This gave dromedary riders much more of a military impact. All this helped the Nabataean Arabs of the city of Palmyra in Syria, to dominate desert trade routes. The Parthians and Sassanian Persians also made use of camel units; the Parthians even experimented (unsuccessfully) with cataphract or heavily armed camels.

Xerxes was one of the next generals to use camels and he enrolled them into the cavalry for frontier duty as well as transport. And so the camel corps became a real feature of Western Asiatic warfare. In 480BC Xerxes took camels with his army to Greece where they were seen for the very first time, and were attacked by lions... as legend has it "out of patriotism".

Gradually the Arabs bred a 'thoroughbred' camel through careful selection for size, speed and strength and these were mainly destined for war. The new way of camel warfare was to have two riders of which one drove the camel and the other was armed with a bow and arrow or with a spear. Another extra was to lead a horse alongside the camel which the second rider might with amazing dexterity mount from the racing camel, for a final attack.

Alexander the Great was never one to miss out on an innovation so after he had sacked the Persian capital of Persepolis in 330BC he supplemented his other baggage animals such as donkeys and mules and cattle with 5,000 camels to carry away the plunder.

Of course in Islam the camel was "The Gift of God" and held in the highest regard together with horses for taking to the Holy Wars that Mohammed had embarked upon. Strangely the camel was never important in Spain, which has in the south rather good desert conditions, not even after the Muslim conquest (711AD).

The Romans maintained a corps of camel warriors to patrol the edge of the desert against nomadic invaders, who were always the scourge of the Empire. The Emperor Trajan definitely had a camel corps in Syria and in the C2nd a Roman camel corps was created under Hadrian and by the C3rd and C4th camels became more and more numerous and as a useful side effect employed for ploughing and pulling carts. The Emperor Claudius even took a few over to Britain, together with his war elephants, to impress and actually terrify the tribes.

'Zamburaks' was the term for soldiers mounted on camels with special swivel guns. 'Zamburak' means little wasp. Another weapon was the 'Shatunal' which means literally 'camel gun barrel'. They were used to great effect at the three Battles of Panipat by Babur which led to the beginning of the Mughal Empire. They had originated in Egypt and were adopted by the Arabs then spread to Persia, India and Afghanistan. They were accompanied by camels bearing drums to make a fearsome noise when attacking and were used against the British in the Afghan and the Anglo Sikh wars. They were employed right into C19th. In the 2nd Afghan war 1878-80, 70,000 camels were lost by Russians and British in the Karakoram Desert.

One European who used the camel in his campaigns was Napoleon who after the Battle of the Pyramids (1798) against some of the Mamelukes, created a camel corps two thousand strong using camels he had bought in Syria.

Another tactic that evolved in desert warfare was forming the Camels into a circle, couching them down and then firing guns over their bodies – a manoeuvre most other animals would not tolerate. In the mid C19th they were used to carry military equipment as well as fighting. But on the whole, until recently, the Westerners were not effective in the use of camels and in fact the English were extremely bad at handling them. They just couldn't seem to understand how to manage these exotic animals, they either over worked them or underfed and under watered them, and in the Great Game when fighting in the 2nd Anglo-Afghan war (1878 – 1880) they lost over 70,000. Another example was that at the relief of Gordon at Khartoum (1885) the camels were still being treated badly and stupidly.

Jefferson Davis wanted to employ camels to pursue the Comanche Indians because although horses are faster, camels have greater endurance in almost desert conditions. So in 1855 he imported seventy odd camels from Turkey and Egypt but they were not a success as not only did they panic all other animals but the grooms who came over with them turned out to be totally ignorant of their charges.

Not until the C20th did camel management become more efficient and more humane. The French enlisted General de Lay's Sahara Camel Corps, which was made up mainly of the Méharistes, the toughest, most hard bitten men that North Africa had to offer and which formed part of the Armée d'Afrique to patrol the boundaries. The desert Bedouin tribe the Chaanba of North Sahara manned the unit with French officers in full hussar dress. They may have made a most romantic sight but they were also extremely tough and effective and were the most feared troops in the desert war. They were likewise used to 'pacify' traditional enemies such as the Tuaregs. This they took a great delight in doing, as the tribes had always conducted raids or *razzias* on each other. They had much more trouble with the Reguibat and the Maures, who simply melted away into the Spanish Sahara when there was trouble. Another important contribution the Méharistes made was to the European knowledge or rather lack of knowledge of camels and the desert and how to survive in it. The French were particularly good at fraternizing with the nomads and had learned their dialects and come to admire their habits and culture. The last camel corps was only disbanded in 1962 with regret by both cultures.

The Bikaner Camel Corps was a military unit formed by Maharajah Ganga Singh in India for use by the British on various fronts. There was a long tradition in the Rajah's family reaching back to the mid C15th. The corps was used against the Chinese in the Boxer Rebellion of 1900, the Somali Uprising 1902-1904 and in Egypt in the WWI and even in the Middle East in WWII. There were problems moving camels to these distant places as they are prone to sea sickness and airsickness.

In the 1st World War, the camel had its moment of glory when Lawrence of Arabia led the Arab Legion on his legendary camel charge on Aqaba. This Turkish stronghold at the top of the Red Sea had all its defences aimed at the sea so Lawrence stole a march on them by mounting his attack at night across the desert and coming in from behind their guns.

During the Algerian rebellion (1954) camels became essential once more. They had only partially been replaced by motor transport, which was sometimes used to transfer them from place to place but for crossing desert camels are incomparable.

Osama Bin Laden had a more modern use for camels. In the early 1990's he gave huge consignments of Kalishnikovs to 'Islamic Jihad' which were sent on camel trains across the border into Egypt. Some of these were to be used in the attack on Mubarak.

Now for their cousins the llamas! Llamas had been domesticated at least 2,000 years before and as the only beasts of burden they were capable of carrying no more than 80-100lbs each and could travel 15 miles a day. But they are unable to pull a plough and even more essentially cannot be ridden. So they were useless against the invading Spaniards.

The next of our animals the Reindeer made an appearance in World War II when about 10,000 of them were used in Finland by the military for raiding patrols, for supplying transport and for taking the wounded back to field hospitals. The Soviet Army established the only Reindeer

Transport Battalion which consisted of 6,000 reindeer and about 1,000 herders. They were involved in the only major military operation that has taken place in the Arctic in modern times. They transported ammunition, food, the wounded, including pilots who had survived a crash. They were also used for towing damaged planes to be repaired. It was largely thanks to them that parts of Norway were liberated from occupying Germans.

Next and predictably the crucial question was how could the elephant be used in warfare? The main destination man wanted to ride his elephants to, was battle. And at first they seemed to be a most potent new secret weapon. The psychological advantage was tremendous – gigantic and bizarre creatures bearing down on infantry and cavalry alike – they must have seemed overwhelming. A charge of elephants must have struck the same awe as the Panzer divisions in World War II rolling across Europe, or the first tanks going into battle at the Somme in WWI. The first mention of the military application of elephants, found in Sanskrit hymns from 1,100BC tells us that the very sight and smell of them could panic man and horse alike. The thickness of their hide was another benefit. Armoured elephants were used as spearheads leading horses and chariots into attack by Aryan generals. In India they even armed the trunks with swords and the tusks with poison and trained the elephants to attack both men and horses. Sometimes they trampled on the soldiers and sometimes they picked them up in their trunks and passed them back to their mahouts to deal with.

But then once man had got over the first shock of the new secret weapon he began, as always, to find ways to defend himself and counter attack. At first, panic was all on the enemy side but then it was discovered that elephants too could be terrorised. One of the earliest anti elephant device, as forecast by Pliny the Elder, was the use of pigs. During a Peloponnesian War a siege of Megara (C5th BC) was broken when the resourceful Megarians poured oil on a herd of pigs and set them alight to run through the elephants. What a scene that must have been! Smoke from the flaming oil; the noise of the burning pigs squealing and the reek of roasting pork; then even more chaos as the trumpeting elephants panicked, turned and ran riot through their own screaming troops.

There were numerous battles with elephants fought during the last three centuries BC in Asia. One of the earliest of these was the Battle of

Gaugamela (331BC). Darius and the Persians had 30,000 men and for the first time employed elephants against Europeans, they used fifteen of them. The Persian elephants had made such an impression on the Macedonians that Alexander had felt it necessary to make a sacrifice to the God of Fear the night before the battle. His tactics the next day was firstly to keep his cavalry well away from the elephants and then he made use of caltrops, these are five-spiked plants in the wild and when copied in iron have a lethal effect as some spikes are always upper most. As a result Alexander won the battle even with only 7,000 men. Then five years later, at the Battle of Hydaspes (326BC) having learned from the Battle of Gaugemela one way of dealing with elephants he then went on to employ a particularly vicious method against the enemy's elephant cavalry, he used eighteen-foot pikes tipped with curved cleavers to cut into their trunks. Nevertheless the elephants' performances had managed to impress him so much that he decided to incorporate them into his own army.

The Emperor Akbar of India (304BC) is supposed to have been the greatest elephant rider ever. Some of the emperors had chariots pulled by elephants and their war elephants were numbered in thousands and their horse cavalry in the tens of thousands. The rulers were known as the Lord of the Elephants, and had tribute paid to them in elephants. The Indian Moghul Emperor Jehangir in the early C16[th] had no less than 40,000 elephants in his stables. The feeding of them must have been an ongoing chore as elephants being giant vegetarians require over 600 lbs of leaves and grass a day.

In the following centuries, further use of war elephants in Europe was mainly against the Roman Republic. Elephants were even employed in Italy by King Pyrrhus of Epirus at the Battle of Heraclea (280BC) in the Macedonian wars. He had gone to the help of those Greek colonies, established in the instep of Italy, that were being threatened by Rome. This culminated in the first clash between the Roman legions and the Greek phalanxes. And at first they were pretty evenly matched, but then the Greeks mistakenly thought Pyrrhus had been killed as he had changed armour with one of his body guards and the Romans were in the ascendancy. However Pyrrhus reappeared without armour to galvanize his army. He had also kept back twenty war elephants, lent by Ptolemy II, in reserve topped with towers filled with soldiers. When he threw them into the fray victory was theirs, but at such a heavy cost it gave rise to the phrase 'Pyrrhic Victory'

The 3rd Ptolemy, Euergetes (246-222) fought many successful wars against his Asiatic rivals with both sides using elephants. He not only used Indian elephants but also African ones from North Africa, where they are now extinct. This was probably how the Carthaginians came by their elephants and also imported mahouts from India to train them.

By far the most notorious march was during the Second Punic war (218BC) when Hannibal having marched through Spain and Southern Gaul, then crossed the Alps to invade Italy with thirty-seven elephants. The elephants were amazingly sure-footed and safely carried the Carthaginians across the frozen Alpine passes. Each elephant was driven by a mahout and carried three heavily armoured men in a *howdah*. Cato the Elder, in his book "Origines", instead of flattering the vanity of the nobles only gave praise to the bravest elephant in battle, Surus (the Syrian).

The successful military use of elephants spread across the world. The successors to Alexander's empire, the Diadochi, used hundreds of Indian elephants, probably mostly the larger ones from Sri Lanka, in their numerous wars. One of Alexander's successors was Selleucus Nicator, (358-280) who held northern India until Chandragupta decisively beat him with an army containing 9,000 elephants. This was the start of the first unifying of India by Chandragupta and the beginning of the Mauryan Empire. The Egyptians and the Carthaginians began taming African elephants for the same purpose, while the Numidians used the Forest elephant. The African savannah elephant, larger than the African forest elephant or the Asian elephant, proved too difficult to tame for war purposes and was never widely used.

Another battle involving elephants at about the same time was the Battle of Raphia (217BC) fought between Antiochus III and the Seleucids of Syria using the fairly large Syrian elephants of which he had 102, against Ptolemy IV of Egypt who had 73 of the smaller African forest elephant. Both sides used their elephants in the same way – to screen and break the formation of the charging cavalry. However Ptolemy's elephants were better trained and several sources maintain that the Seleucid elephants panicked and fled the battlefield causing the usual carnage with their stampede.

In Hannibal's last battle Zama (202 BC), near Tunis, his charge by eighty elephants was largely ineffective, because Scipio had ordered the Roman maniples (tactical units) to open up corridors for them to pass through the lines. Horns were sounded to scare the charging elephants and Hannibal was defeated, putting Rome back as the head nation of the Mediterranean.

More than a century later, in the Battle of Thapsus (46 BC), Julius Caesar used another ploy by arming his fifth legion (Alaudae) with axes and commanding them to strike out at the elephant's legs. The legion withstood the charge and the elephant became its symbol.

Now moving to C12th, during the war the Khmer army waged against Champa, a Hindu kingdom situated in modern Vietnam, the defending Chams invented a two-man crossbow to be carried on the back of the elephant. This combination acted like a mobile artillery unit.

The great anecdotal Marco Polo, during his famous travels in Asia at the end of the C13th, describes fighting between Mongols and Burmese. The former had cavalry numbering 12,000 and the Burmese had 60,000 men plus 2,000 war elephants. The horses, as so often the case, refused to go anywhere near the elephants and so the Mongol general used another ploy. He had the horses tied up and then the soldiers, from behind a covering of trees, fired volley after volley of arrows at the elephants, until they too refused to be used in battle and finally victory went to the Mongols.

Timur (or Tamerlane) practiced another method of fending off elephants in battle in 1398, when he was trying to sack Delhi. The Indian Sultanate had struck terror into the hearts of the Turks with the deployment of 100 Indian elephants but Timor had set straw alight on the backs of his camels before the charge. When the smoke made the camels rush forward, the tables were turned and the stampeding elephants retreated and again they too crushed their own troops. Timur was also the first to use deep ditches as efficient barriers to the elephants which are, of course, incapable of jumping.

In 1526 the Battle of Panipat, already mentioned in connection with horses, was particularly memorable not only for this but also for the use of gunpowder and firearms for the very first time in a battle in India. Ibrahim Lodhi's army was double the size of Babur's and he was fielding 100 war elephants but he had no guns. Babur had the guns; so not surprisingly, Lodhi's elephants, terrified by the sound of the gunfire fled the field, trampling their own side. This was the beginning of the end of elephants in battle.

Both sides, the Mughals and the Rajputs, used elephants in the Battle of Haldighati, which took place in 1576. Pratap Singh, the Rajput had clad his famous horse Chetak in armour fashioned to resemble a grotesque elephant, working on the assumption that this would not only terrify the Mughal horses but counted on the assumption that the enemy elephants would not injure what they saw as a young calf.

The last known use of elephants in war was by the Iranians in 1987, who used them to transport heavy weaponry for use at Kirkuk. But with the coming of gunpowder the elephant, that most unreliable of all weaponry had fought its last battle. Other legends are, not surprisingly, of elephants going to war as part of the four arms of an army – chariots, cavalry, foot soldiers and elephants.

So elephants were no longer used in actual battle although they did come into their own for transporting military equipment across awkward terrain. For example in the last War, elephants were invaluable on the Indian and Burmese frontier, not merely as a means of transport, but also as a way of building bridges and roads, hauling guns and engineering equipment over mountains and through jungle – areas where it was impossible to get even tractors in, much less use them. Finally, there is a story that during the Burma campaign, a British Mule team carrying supplies was actually charged by Japanese riding elephants.

So, all our six mammals and even reindeer had their uses in war. Now with modern weaponry you would think they could be retired, but sniffer dogs, transport in difficult conditions, and special uses still demand their services if to a far lesser degree.

ET CETERA

1. To go from the supremely important to the sublimely ridiculous – there is a rather amusing story from mediaeval times concerning the Battle of Sempach in Switzerland (1386). The fashion for knights at that time was very long pointed toes on their armour

to stop their feet slipping out of their stirrups. (Heels had not yet been invented). But when the knights dismounted for the obligatory hand-to-hand fighting, they found that they had to cut off several inches of armoured toes before they could even walk, much less fight. A lovely thought – a battlefield littered with severed armoured toes!

2. Another light-hearted legend has come down regarding a battle between the Crotonians and the Sybarites, in the C6th BC. The Sybarites, living up to their name, had taught their horses to dance to music. The Crotonians, knowing this eccentricity, played dance music before the cavalries joined battle, with the expected ridiculous outcome of dancing enemy steeds.

3. In WWI fox hunting was considered such good training for officers that a day's hunting was not counted as part of a soldier's leave.

4. In WWII Smokey, a 4lb Yorkshire terrier which belonged to Corporal Bill Wynn, flew for 2 years on combat missions in fighter planes and became a famous mascot.

CHAPTER 13

UNINTENDED CONSEQUENCES

Dog (Byron) from Canine Partners, Midhurst, Sussex. UK

As we have seen animals became attached to us partly by accident and partly by design and with each and every one of them unexpected results and unintended consequences developed down the years. We took them on for hunting, for transport; for pulling ploughs and carts, for war but then our relationships went on to develop in so many unanticipated ways. They came bearing gifts, unimagined when they first entered our lives. We have reciprocated sometimes with dire results.

Starting with man's best friend, dogs are not only the most popular pets in the western world; they perform many jobs that only they can achieve. For example their sense of smell is so strong that it can detect one part in a quadrillion and perform very much better than a modern odour-detecting machine which only detects one part per billion. The reason for this talent is that dogs have 220 million smell sensitive cells or olfactory receptors over an area the size of a pocket handkerchief, whereas we humans have a mere 5 to 10 million over an area the size of a postage stamp, making their ability to identify a scent forty times better than ours. So this makes them of great help to the police and customs officers in guarding, tracking, sniffing out dangerous substances, dead bodies, arms, in town or countryside. They are also used in prisons to check up on visitors or even the odd crook warden trying to smuggle illegal things such as drugs into the prisoners.

Eric Burchell runs a detective agency with a difference...ACT Ltd ..."Advanced Canine Technology" is staffed by 'Sniffer Dogs' that travel the world to sniff out undesirable substances such as narcotics, explosives, and illegal animal parts. They can also be trained to smell out smuggled ivory, and with the use of DNA even the actual origin of the tusk can be ascertained. Another British organization called FACT (Federation against Copyright Theft) employ two black Labradors called Lucky and Flo which are expert at detecting contraband DVDs, which apparently have a very strong smell. Dogs can be trained to home in on a particular scent, and become a specialist in that smell be it cocaine, rhino horn or elephant ivory. They are essential in modern times as smuggling has gone up recently 800%. Many of these illegal things are shipped out by the Chinese. In Democratic Congo CITES is completely ignored and tusks are sold openly in Kinshasa.

The sniffer dogs are also expert at finding drugs sometimes hidden in bizarre places such as in the wheels of roller suitcases or the drive shafts of cars. Another even more specialized use is in California with dogs who can actually detect the invasive Quagga mussel from the Dnieper River in the Ukraine, which can attach itself to the underside of visiting boats and become invasive. They have also been trained in the USA to sniff out infestations of termites before they can do too much damage to houses.

Another unexpected use is for dogs to find bumblebee nests which apparently have a very strong smell. Bees of course are essential for pollinating our crops and badgers are responsible for eating a large number of their nests.

Belgium was ahead of everyone in modern Europe in training dogs for special duties and by WWI German Shepherd dogs were being employed for military duties and then moved on into police duties in the 1920's and 30's. Finally England got the message and in 1938 two Labradors were based in South London, but of course WWII was fast approaching and not until after that was a small training school founded and the German Shepherd became the main dog of choice. They each cost about £6,000 to train not only to use their ability to scent but also to chase and attack and help control individuals even in crowds and in the face of weapons. The Defence Animal Centre (DAC) was opened in 1946 in Melton Mowbray (Leicestershire) to train dogs for all three armed forces. They also train dogs for the Police service, the customs, and the Prison service. The very sight of the dogs with their handlers seems to act as a deterrent to some criminals. In London 1,400 crimes are reported every night and the dogs can even smell out firearms that have any recent human scent on them and also cash stashed away.

As shown dogs can be trained for the tough jobs of war and policing but the next phase dogs moved into was as carers for blind and disabled people and even handicapped children. This can easily be imagined as starting informally as far back as when Man and Wolf first combined. There are now many organizations dealing with training these highly specialized dogs and it can cost as much as £10,000 and 170 hours to train a support dog, with no charge for the fortunate recipient. Dogs for the blind have been invaluable since the C16th but in modern times the first training schools were started up in Germany to deal with the men blinded in WWI.

One of the famous is Buddy, a German Shepherd bitch who became the guide dog to Morris Frank. The story is a marvelous one. Morris lived in Nashville, aged 20 and blind which made him very depressed because he could not be at all independent. In 1927 his father read him an article about a woman called Dorothy Eustis who was training dogs in Switzerland to guide blind people. Morris and Dorothy met in Switzerland

and Morris acquired Buddy. On return to America they set up "The Seeing Eye" the first training school for the blind and dogs in the USA. Buddy became a national hero and today the school has trained over 14,000 dogs.

Now all over the western world blind people with guide dogs are an everyday sight but there can be awkwardness with Muslim taxi drivers in London who sometimes refuse to have them in their taxis, in spite of the Sharia Council of Great Britain decreeing that an exception should be made for guide dogs.

"Hounds for Heroes" have their headquarters in Essex and have been set up recently to help men crippled in warfare. Another charity "Canine Partners" in West Sussex mainly uses the special cross breed the Labradoodle, which combines the intelligence of the poodle with the good nature of the Labrador. The charity trains them to be helpmeets for the disabled, for which duty they can obey over one hundred words of command and can even pick out shopping in a supermarket, and put washing into a machine. The disabled person comes to stay for two weeks so a suitable dog can be selected. Both the person and the dog need training together before going home.

Moving into the medical world, it was found that dogs could detect the onset of an epileptic fit and are specially trained at the cost of about £10,000 by "Support Dogs", a charity based in Sheffield, UK. The warning can be anything from 25 minutes to 45 minutes but is always constant between an individual dog and owner which enables the patient to be able to make arrangements to get into a safe place before the seizure takes over. Dogs with their keen sense of smell can even detect diabetic sugar levels and bladder cancer by sniffing a urine sample. Labradors and Portuguese Water dogs can apparently detect lung and breast cancer simply by smelling the patient's breath. The NHS is now testing out dogs, trained by Medical Detection Centre, to detect cancer. They are starting with prostate cancer, the most common form of cancer in men in the UK. A nursing home in Ohio USA has a dog that senses when a resident is due to die and will stay with the patient until the nurses or family take over.

A 2 year old beagle has been trained to sniff out *C.difficile* in two hospitals in Amsterdam. When he detects the smell he lies down near the patient. He has an 80% success rate and can sweep an entire ward in ten minutes. Conventional diagnostic tests are much slower and of course more expensive. *C.difficile* can be fatal and early detection can prevent the spread.

In other spheres dogs and horses have both been found to have the most amazing curative powers with regard to humans. Children with all varieties of mental states seem to be reached by them and this can have the most humanising effect on them. PAWS standing for 'Parents Autism Workshop and Support' help children who suffer from Autism, Asperger's, Spina Bifida, Cerebral Palsy and even Down's syndrome. In Australia in 1996 'Assistance Dogs Australia' (ADA) was started. The dogs, usually Labradors or Golden Retrievers undergo two years training for helping people with disabilities. Dogs are also used as signal dogs for the deaf. Another less obvious present from the dog to its owner is the connection with people who love dogs who stopping to admire the dog end up talking to the often lonely owner.

On a more general level it has definitely been found that people's heart rate and blood pressure improve when they have animals to pet. During the Cuban missile crisis in 1962 President Kennedy attended sessions with his son's Welsh terrier Charlie on his lap to calm him. Freud used dogs to comfort patients and they are used in psychiatric wards to make the break through to withdrawn patients. Even back in Roman times the ladies had lapdogs for comfort and it was thought that their warmth could cure a stomach ache. Recently in America experiments are being carried out in some of their prisons; inmates were placed on a program to help train rescue dogs for adoption, for the blind and for other work. It was started in 1981 by Sister Pauline Quinn in a woman's prison, the Washington State Correctional Centre for Women and it has proved to be an overwhelming success not only for the dogs but for the mental health of the prisoners, many of whom had never had anything to love in their lives before, or to love them. The training of the dogs also helps the prisoners by giving them a purpose in life and often eases states of depression and even violence. The prisoners were taught to rehabilitate unwanted dogs destined to be destroyed and transform them into suitable pets. Some of the rescued dogs have been so traumatized by bad treatment that it can take several weeks to win their confidence and make them acceptable for a new life in a family. The dogs are rescued from high kill animal shelters, neutered and vaccinated; some of them even go on to be trained to help people with disabilities. There are now similar programmes all over USA, Canada and Europe. In 2004 a similar program was started up in Kansas called 'Safe Harbor Prison Dog Programme'. This idea was also taken up in Australia in 2002 by the ADA.

Following an animal scent is another whole game; foxhounds follow the scent on the ground left by the glands on the fox paws, but Bloodhounds can sniff out skin particles and sweat from a human on the ground or floating in the air. Funnily enough when dogs follow poachers in Africa they often lose the trail if their prey crosses a river but elephants, which have a fourteen times better scenting ability even than dogs, will pick up the scent again and consequently are being used more and more in anti-poaching work.

Breeding has perhaps gone too far in the C21st as Charles Darwin observed that "Breeders habitually speak of an animal's organisation as something quite plastic, which they can model almost as they please." We have come a long way – perhaps too far since the first ever dog show was held in 1859 for 60 Pointers and Setters.

In April 1873 twelve Victorian gentlemen started the Kennel Club to try and control all canine matters ranging from dog shows to registration and creating a stud book. And the first stud book was produced by the Kennel Club, giving birth to the concept of the 'pure bred' pedigree. In the UK three quarters of dogs are 'pedigree'. Breeding dogs has continued to 'evolve' as director by Man. Dogs are bred to be pretty, to be sweet, and to have a childlike appeal. Colours and mixtures of colours, smooth coats, curls and waves are selected. If a Rhodesian Ridgeback has a puppy lacking the ridge, it is put down. The bulldog head can be too gross for normal birth and the bitch often has to have a Caesarean. Flat-faced Pekingese and Pugs cannot breathe properly. Some Cavalier King Charles spaniels have brains too big for their skulls, an agonising condition known as 'syringomyelia'. Inbreeding is often used as a quick tool to mould the desirable deformity. Fathers are mated with daughters or even granddaughters, brothers with sisters. For example the 10,000 pugs in UK share the gene pool of only fifty. These 'breeds' are inventions of breeders who have completely lost sight of the fact that all dogs originally are descended from wolves. One third of pedigree dogs are suffering from one of over 500 genetic diseases by the age of five; as a result health insurance for 'mongrels' is less than half. Another reaction is that airlines won't fly pekes or bulldogs because of their breathing problems.

Then, at last a small breeze of sense blew in; the BBC produced an investigative documentary 'Pedigree Dogs Exposed' in 2008 which criticised Crufts and the Kennel Club and condemned the exaggerated breeding objectives that overrode the health of show dogs. The BBC then withdrew all coverage of Crufts and in 2012 it produced a follow up program. The Kennel Club was left with a juggling act between wilfully blind breeders and the health of dogs. A special group called APGAW has been formed (Associate Parliamentary Group for Animal Welfare) which is involved in the health of all animals in the UK.

The other scandal it exposed was that of puppy farms, a misleadingly bland name for using overworked bitches for breeding puppies in the most appalling conditions. To quote from the Kennel Club which declares "Puppy farming is a cruel and abhorrent trade". Buyers of puppies would be wise to follow the Kennel Club's guide lines.

Now luckily there is a fashion for new crossbreeds with marvellous names such as Labradoodles, Springador, Sprockers, Cockerpoo, Jackadoodle and Cockerdoodles. Also it is now acceptable is to give homes to 'Rescue Dogs', mongrels or 'Pavement Specials' as they are known in South Africa. They often make great and grateful pets providing they have not been too traumatised by previous experiences.

In complete contrast, dogs are still being eaten in China and this practice continues strongly with dogs being specially bred for the table as distinct from being kept as pets. Dog meat is sometimes called "fragrant meat" or "mutton of the earth" and is considered an expensive delicacy and to possess special medicinal properties especially in winter. The latest fashion is the importing of St Bernards from Switzerland for the table. They are nick-named "Big Dumb Dog" and are ideal for meat producing as they have huge litters usually once a year and the puppies put on weight so quickly that they can weigh 30lbs within a month and by 4 months they are ready for market. However since 2004 a contrary society called Chinese Animal Protection Network has come out against dogs as meat or being killed as an excuse for rabies control. Dogs are also being brought into families as a substitute second child but as they were killed in their millions during the cultural revolution as being deemed too 'bourgeois' there is no tradition existing of dog ownership and you can see the bizarre practice of dogs being carried when being 'taken for a walk'.

In August 2010 Iran's 'Ministry of Culture and Islamic Guidance' banned all advertisements for pets and pet products. Apparently the old anti-dog culture had been changing and becoming 'westernised' and 'decadent' so a Fatwa citing the old Islamic tradition of anti-dog was issued by no less than the Grand Ayatollah.

For different reasons, in our western domestic pet relationships, cats and caged birds are taking the place of dogs, as being less trouble for lazy or working owners in modern urban cramped living conditions.

Sometimes dogs have become our nannies and supporters as well as our servants – Nana, the dog Nanny in Peter Pan is not so fanciful – but still many of us abuse them and hundreds are put down or even worse abandoned in the streets by people who have tired of them or found them a nuisance or an expense. In Italy there are thousands of feral dogs, who do not breed very successfully but their numbers are replenished by discarded pets. The nations that own the most dogs are the French and the Americans, where every third family owns a dog. In England it is every fifth family, but perhaps that is partly because of the numerous immigrants, many of whom are Muslims. The Swiss and Germans own the least dogs at one dog to every 10th family.

Of course one undesirable consequence of our association with dogs is disease, the most feared being rabies but parasites such as hookworms, tapeworms and fleas and ticks can be transmitted from dogs to us but actually we catch far more deadly diseases from other human beings.

Of all our six mammals, cattle have become important in several specialised ways. Oxen, as draught animals and beasts of burden have all but vanished from the roads of the western world except in Spain where they are still used to draw wagons for their numerous fiestas. But in the developing world, Africa, India and Asia, they are still very much in evidence, ploughing, as beasts of burden, for milk and manure.

In India they are venerated for their sacred powers. Unfortunately the Indian cattle as well as humans are often given a drug for anti-inflammatory purposes and this has turned out to have the most terrible side effect as it is deadly poisonous to vultures. These birds which can resist plague and botulism and rotting fly blown meat are totally unable to take the anti-inflammatory drugs even in the second hand form in the flesh of the consumer of the pills. As a dire result at least 97% of India's vultures have been wiped out. This is bad enough generally but there is another side effect. The Parsee religion place their dead in tall towers called 'Towers of Silence' and the vultures come to do their work and consume the corpses. It is called 'Burial in the Sky', so now there is a long queue of dead Parsees owing to the tragic vulture problem. The drug was banned in 2005 but it will take some time for the reversal of the desperate effects.

But it is in the Western world that the role of cattle has frankly got out of hand in another respect. They have continued to be more and more valued by man for their milk and beef. In most of the world the spread of cattle has developed into an epic of holocaust proportions. As we saw, the habit of eating beef had been taken across to Britain by the Romans when they invaded in AD43 and subsequently the British became the greatest beefeaters in Europe. Even the Yeoman Warders were given the soubriquet of 'Beefeaters' because right up to the early C19th part of their salary was paid with hunks of beef. Then when the English were running out of pastureland they turned to Ireland and forced the disenfranchised Irish to farm on smaller and smaller plots, with their bigger and bigger families. So the Irish in desperation turned to an easy monoculture – potatoes, and the potato blight in the mid C19th caused the terrible and well known famine and exodus. This poisoned relations between England and Ireland and we are still feeling the results today.

The next candidate for providing beef and pasture was North America. The movement of cattle around the world has had great impact on many environments. As we saw Christopher Columbus first introduced them to the New World. One of the main reasons for his voyage to find a new route to the Far East was to bring back the spices that are used to preserve beef. The breed he took with him was the Spanish Longhorn, a very tough animal which was highly adaptive and became browsers as well as grazers. The herds grew and grew into millions and the new owners, the Beef Barons became the wealthy elite. There were just two awkward blocks to untold wealth - the huge herds of Bison and the Amerindians who depended on them. But that was solved by the most horrific wholesale slaughter of a single species by man the world has ever seen. The buffalo which had been there for at least fifteen thousand years vanished almost over night – over 4 million of them. The Indians couldn't believe that the vast herds on which they always depended had vanished; they never really recovered from the shock – their world was shattered. Most of the dead buffaloes were left to rot, but some bones were collected for fertilizer and some horns for buttons, glue and combs. This carnage made room for the imported cattle. The new herds grew and grew.

A mass of fairy stories or folk legends accrued about the frontier being pushed back with an almost religious drive and of man's dominion over beast. The Cowboy has achieved almost mythical status in American culture. Films galore have been made in which they are invariably portrayed as dashing heroic and very macho men. The truth is that they were usually poor landless men, certainly in Chile where they were known as 'hauso' or in the Argentine 'gaucho', 'llanero' in Venezuela, 'vaquero' in Mexico. The growing herds of cattle were often owned by absentees – British financiers and aristocrats who were greedy for beef, the ultimate luxury of the upper class diet and a great and growing investment.

Reasons for eating Beef? Well apart from being energizing delicious food, it signals wealth, status, and machismo. It indicates you have arrived in the top billion of those people who can afford to overeat, if they have a mind to.

The next problem was to get them to market or rather the Chicago abattoirs. In 1864 Jesse Chisholm opened a wagon trail which became known as the Chisholm Trail, along which passed millions of cattle from Texas to Abilene to catch the trains onwards to Chicago. Refrigeration had also been invented by then and from 1875 onwards carcasses were shipped to England. To fatten them up cattle were being increasingly fed on grain, which is not their natural food and can upset their digestive systems. An astonishing 80% of US cereals go to feed animals, mainly cattle. This diet, as opposed to the natural grass, puts on weight and fat, which is passed on to the human consumers. To enclose the vast land seizures a new simple but effective invention by Joseph Glidden (1874) was employed – barbed wire, nicknamed 'the devil's hatband' by the cowboys.

The Americans are now the greatest beef addicts and the fast food outlets such as McDonalds and Burger King provide hamburgers for a population most of whom eat out at least once a day. The world of the obese had arrived.

Meanwhile the world population is forever growing and growing and as more people moved up the meat ladder even the vast prairies were not big enough to satisfy the hunger for beef. So the next continent ripe for plunder for 'free grass' was South America. In Brazil, Mexico and the Argentine, cattle were at first kept by the Jesuits at their missions. Here the barriers were the Rain Forests and the rural indigenous population. Money seemed to be no object so they cut down the forests and ignored the plight of the peasants. Thousands of square miles have been cleared to produce land poorly suited for grazing and which after two or three years, without fertilizers becomes useless and it is necessary to cut down yet more forest. Apart from the local disasters of pauperizing the rural peasants and depleting and poisoning the water, the loss to the world of the tropical eco system with its amazing range of animal and plant species is beyond calculation. Just to look at it from the narrowest selfish human point of view, more than a quarter of all medical and pharmaceutical products are derived from tropical plants and three quarters of cancer drugs alone. Who knows what marvellous cures may be lurking still to be discovered – maybe even one to help counteract human stupidity! Of which Einstein would have approved.

In Brazil alone 81% of the land is owned by 4.5% of the population. Operation Amazonia was launched to convert forest into grazing land. In this last great land grab they were helped financially by the World Bank and the United States Agency for International Development (USAID). The greed for beef has reached enormous proportions; over 48 million metric tons of beef were consumed globally in 2000. Argentina consumes the most beef per capita, even more than the United States, Australia, or Canada.

There are now 1.3 billion cattle in the world and over a quarter of the landmass is used for pasture, and cattle are consuming enough grain... about a third of the world's harvest, to feed hundreds and thousands of people. The highest proportion numbering 180 million is in India alone, but these cattle are, of course, not used for beef. Brazil and China have about 150 million cattle each and the USA has over 100 million. More and more forests are being felled and more and more land is being so overgrazed that desertification is increasing on a grand scale especially in South America and Africa. But beef and hamburgers are the hallmark of having arrived out of poverty and the poor of India and China and the rest of the world aspire to having this status symbol on their menus. China alone has increased its consumption of beef by nearly 300% in the past few years.

One thing is certain – we are mucking up the planet. Apart from forests disappearing at a terrifying rate; oceans are being poisoned by the run off of fertilizers to grow the food for cattle which are ever increasing in numbers to keep up with feeding the human population which is also relentlessly increasing in numbers. Another side effect is that Cattle give off methane gas which is 23 times worse for the atmosphere than CO2. They are responsible for 18% of greenhouse gasses, which is more than all man made forms of transport – planes, cars and lorries put together. What are we doing?

Starvation is one end of the scale but at the other end, Americans and Europeans of the industrial world are dying from the diseases of affluence – heart attacks, obesity, diabetes and cancer – often the result of eating too much beef. What is the main dish offered in the millions of fast food outlets? Hamburgers – ground BEEF.

In Africa cattle were always venerated by the Africans but their herds were small and often nomadic. But even here there is trouble. The nomadic life was discouraged by arbitrary fencing and colonial disapproval through the administrators of government services. Wells were drilled to encourage settlement, the wear and tear of cattle going to the water holes and overgrazing on a grand scale resulted in desertification.

Another place they were introduced to was Australia in the 1820s and 30s. Later they were set free and became very successful (for them) feral animals, adding to the list of animals that Australia, too late, wishes had not been launched into her countryside.

India also has a cattle problem of a completely different kind. When visiting India one of the things that strike you is the sight of cattle wandering freely in the streets and the elaborate care everyone takes not to injure them. In Kashmir until recently, killing a cow, even accidentally, carried the death penalty and even now it could result in life imprisonment. Hindus believe that every cow is inhabited by millions of gods. So the difference in attitudes between Hindus and Muslims has sometimes led to cattle riots. The Indian Government actually provides homes for half a million cattle too old and frail to survive on the streets.

Our next animal, horses also play their part in helping disabled and mentally impaired people. This form of treatment is called Hippotherapy. This ability was recognised as far back as 600BC by Orbasis of Lydia. Hippocrates (460-377BC) used horses for various kinds of treatment, then travelling down the centuries to Thomas Sydenham (1624 -1689) the English Hippocrates

In the USA miniature horses are being trained for the same work as dogs for the blind. The Guide Horse Foundation trains them to ride in vehicles, go up and down escalators and even to be house trained. Like in the cavalry and police, they also have to be trained to cope with noise and trouble, and an added bonus, the Muslims would have no problem with them. To deal with slippery surfaces special horse trainers or sneakers have been invented with lace ups very like the ones for people.

Other help for deprived and disabled children comes from the Ebony Horse Club which actually has no horses or stables of its own, but uses other people's. It was started by Ros Spearing in 1996 in Lambeth and Brixton and children with problems are referred to them by schools, the police and also social workers. All the young riders are taught to think of their horse first and a whole new world is opened up for them. When taught to ride horses they get fitter, begin to build muscles and more importantly begin to connect with the world and other people.

Even in modern times donkeys have not been entirely phased out and replaced by motorised transport. There are estimated to be 44 million donkeys in the world, 11 million alone in China. All over the Middle East they are still very much in use. And in Cyprus, for example, only donkeys can manage to work the steeply terraced vineyards of Mount Alphemi. And in Cuba the deteriorating economy means spare parts and even fuel for motors is in short supply. So both transport and farms have become once again very dependent on donkeys and horses.

Draught horses had been the mainstay of transport and strangely the railway in England was very dependent on horses for shunting rolling stock, and the last horse retired in only 1967, just before the last steam engine was shut down. So many horses were in use and so many were being ill-treated that in 1822 the RSPCA passed an Act for the Prevention of Cruelty to Animals.

Another important use of both horses and dogs in England is with the police. Training horses for the police is not all that different to training them for war. They have to learn to cope with extreme noise and violence – instead of camels and elephants and men with spears there are noisy internal combustion engines and unruly mobs. The first mounted police were the Bow Street Horse Patrol formed in 1758. They now use a strong hunter type, which are trained in a depot in Thames Ditton for about forty weeks; their working life is about 14 years. The Mounted Branch stands at about 120 strong with 140 Officers. One great advantage in situations such as football matches is that with a view from eight feet above the crowds the Officer can spot any trouble brewing up. It is estimated that a trained Mounted Officer on a trained horse can be as effective as a dozen officers on foot in such situations.

Not only dogs but now mustangs – the wild horses of America are trained by prisoners to become saddle horses, a process that takes a good four months and turns a $125 horse into one worth at least $1,000 and more to the point also gives a direction to the life of the prisoner that domesticates it.

How is the elephant faring? In Asia it has been largely marginalised by machinery for transport and for work purposes, although it still does logging and other work in Myanmar where it is luckily still vital to the economy and is managed by the MTE – Myanmar Timber Enterprise. Some remote villages have an odd job elephant living in it to perform local tasks. Mahouts in Myanmar are called Oozies and perhaps because of their Buddhist background they are amongst the best in their treatment of their charges. But many elephants are on the brink of starvation owing to logging being on the decrease, the main reason being drastic deforestation. An estimate of numbers taken in June 2003 found that Asian elephant numbers had fallen to below 30,000 and that this population is scattered across about nine Asian countries from China and Thailand to Nepal and India down into India and Sri Lanka. Their migratory routes have been cut off by the ever expanding human population (3% per annum) and their settlements, leading to inbreeding and food shortages. The good news is that the Wild Life Trust of India and the World has opened up a corridor linking two reserves in the Indian Myanmar regions. In Bangkok they had over three hundred elephants and their mahouts begging in the streets, then, in 2003 ninety-nine of them were taken on a pilot scheme, back into the forests to patrol against poachers and illegal loggers. This has proved to be a great success all round.

In 1997 the Elephant Trade Information System (ETIS) was set up at a conference in Harare, 'the capital of Zimbabwe', whose aim was to track the illegal trade in ivory and other elephant parts, with the approval of CITES which stands for the Convention on International Trade in Endangered Species of Wild Fauna and Flora. In Africa two things became vital – habitat protection and preventing poaching. In 1989 the western countries in Europe, USA, Australia and others banned the commercial import of ivory. But poaching is once again on the increase. In the USA Ringling Brothers Circus claim to be running a great conservation, breeding and training scheme for their elephants but the reality is that they have been found guilty of excessive animal cruelty on many occasions.

Now for the African elephant – an estimate of the population taken in 2003 found them to number about 500,000. This is a terrible loss from an estimate of 1.3 million taken in 1979, and 5 million at the beginning of the C20th. Africa had lost over half its elephants to poachers and of course far more males were killed for their tusks. An interesting example of astonishingly rapid evolution is that males are now being bred with much smaller and therefore less tempting tusks. The Chinese government is responsible for over 90% of the illegal black market in the ivory trade. They are also being killed for bush meat which when smoked sells for a lot of money.

However there is a glimmer of brightness on the horizon. Tourism is proving to be a new lifeline for elephant conservation not only in India but in Africa too where they are now being used on safaris in Kenya as well being the star attraction in the 'Must See Big Five' list for visitors to game reserves.

Our next animal the Reindeer, while still essential for milk and meat and its other products are in retreat as so are the Sami, especially in Russia where their lives have been ruined by 'collectivisation' and the Russian passion for 'rationalization' imposed on them. Before borders were mapped out and enforced the Sami ranged across Sweden, Norway, Finland and W.

Russia – 'Fennoscandia' as so did the reindeer. The Russian Sami have been forced into soulless concrete blocks in a new settlement called Lovozero. Hunting reindeer was forbidden at first and then the Russians themselves slaughtered hundreds and thousands of reindeer and elk from helicopters. Then Murmansk was founded on the Kola Gulf which although North of the Arctic Circle and the port of Archangel is free of ice thanks to the Gulf Stream. A shoddy railway was built in about 1914 and the town was a squalid mess of ill-built concrete blocks and peopled with dropouts. In the 1940s the nearby mining for apatite (phosphate minerals) and phosphate for fertilizers, gold and nickel and the processing of the latter led to such bad air pollution that children even in Norway and Finland were suffering brain damage. Then to make matters worse nuclear waste was being dumped. All this of course had an adverse effect of tragic proportions on the reindeer herds and their age old way of life. Even in Scandinavia where the reindeer are luckily still essential for food and milk, their use as transport is being threatened by the quad bike in summer and the snowmobile in winter.

Cars, aeroplanes, high rise buildings, women going out to work as well as their men, increasing urbanisation all are ousting animals from our lives whether as workers or companions. All this has happened in just two hundred years particularly in the Western world and in these few years we have gone from close associations with the six most important animals in our lives to a state bordering on divorce.

Trying to do an overall review of the place of animals in modern times is impossible because luckily for some of these animals not the whole world is living in the C21st. Visit parts of the Muslim world and you are back in mediaeval times or even as far back as the C7th, and in some remote islands and the far north the Stone Age is only just back round the last corner. Much of the deep south of America is embedded in the pre-Darwinian early C19th, and that's not counting the splendidly eccentric Amish, who still use horses for ploughing. The Inuits in the far North of Canada who with their dog sleighs and hunting have led lives isolated from the modern world until the 1940's are now being dragged into modern life because of global politics and global warming.

Transport by cattle, horses, mules and donkeys is still essential in parts of the third world and in the desert the camel can move over terrain that even a four wheel drive cannot manage, and so they are still irreplaceable in the Sahara and parts of Arabia. Camels as noted in Chapter 7 were introduced into Australia for building the railway. There are now rumoured to be over a million feral camels, far too many to be sustained on Australia's desperate desert conditions, but what to do with them? The meat is good and almost fatless and there is a drive to get some of the young male aboriginals to go hunting them.

Clearly Man's attitude to these six animals ranged from priest to hunter to butcher, to trainer, to breeder, to friend and many other roles. In our imaginations, we even made bizarre hybrids to worship that sometimes incorporated human parts, such as centaurs and the Minotaur. We have loved them, used them, feared them, eaten them but the most dangerous reactions may be those of modern western urban man – ignorance and indifference. These attitudes plus Man's overpopulation is sounding the death knell of so many species, including our lifetime friends, and maybe even ourselves.

Perhaps we should end up with a quote from George Orwell's Animal Farm; 'Four legs good, two legs bad.'

ET CETERA

1. Dog Numbers: - 400 million dogs in the world, 60 million in the USA, (but only 10,000 wolves) 30 million in Brazil, 8 million in France and 7 million in Britain.
2. Cattle Numbers: - 1.5 billion. 200 million in India alone.
3. Horse Numbers:- 65 million and 300 breeds.
4. Donkey Numbers:- 44 million
5. Camel Numbers:- 15 million Dromedaries include over 1 million feral in Australia. 1.4 million Bactrian camels.
6. Reindeer Numbers:- Seem to be fluctuating wildly. Caribou 5 million.
7. Elephant Numbers;- Asian is below 30,000. African 500,000 (in 2003)
8. Human numbers:- Nearly 7 billion and increasing rapidly.
9. **Since 1600AD more than 50 mammals and at least 100 birds have become extinct as a direct result of human activities.**

---ooO0Ooo---

BIBLIOGRAPHY

Amirsadeghi, Hossein	"The Arabian Horse"
Barloy, J.J.	"Man & Animals" (London Gordon & Cremonesi 1974)
Bokonyi	"History of Domestic Animals"
Bondarenko Nina	"Hearts, Minds and Paws"
Brandel, Fernand	"Civilisation and Capitalism" C15th – C18th I, II, III
Brewers'	"Dictionary of Phrase and Fable"
Bronowski, J.	"The Ascent of Man"
Budianski, Stephen	"The Nature of Horses"
Budianski's	"The Truth About Dogs" Why animals chose us. New York, (William Morrow)
Bulpin, V.	"Discovering South Africa"
Bulliet, Richard W.,	"The Camel and the Wheel" Cambridge, Mass., 1975
Cansdale, George	"Animals of Bible Lands"
Chesterton, G.K.	"The Ballad of the White Horse"
Clutton-Brock, Juliet	"A Natural History of Domesticated Mammals" 1987
Clutton-Brock, Juliet	"Horse Power" Natural History Museum Publications 1992
Cunliffe, Barry (edited by)	"Prehistoric Europe"
Conrad, Jack, Randolph	"The Horn & the Sword."
Darvill Richard V.S.	"Darvill on Training the English Race Horse – in a Series of Rough Notes." (1846)
Darwin, Charles	"The Voyage of the Beagle"
Darwin, Charles	"The Origin of Species" 20
Dawkins Richard	"The Greatest Show on Earth".

Dembeck, Herman	"Animals and Men"
Demakakoplou etc.	"Gods and Heroes of the European Bronze Age"
Diamond, Jared	"Guns, Germs and Steel."
Dugmore, A. Radclyffe	"The Romance of the Newfoundland Caribou"
Eltringham, S.K.	"Elephants"
Epstein, H.	"Domestic Animals of Nepal"
Evans, J.O. & D.J. Atkins	"The Camel Handbook"
Foster, William edited	"Early Travels in India 1583 – 1619"
Freeman, Michael	"Transport in the Industrial Revolution" 30
Gansdale	"Animals of Bible Lands"
Gauthier-Pilters, Hilde and Dagg, Anne Innis	"The Camel"
Gourand, Jean-Louis	"Horses"
Graves, Robert	"Greek Myths"
Groves, Colin P.	"Horses, Asses and Zebras."
Grzimek	"Animal Life Encyclopaedia"
Harris, Marvyn	"Cows, Pigs, Wars and Witches"
Harris, Marvyn	"The Sacred Cow & the Abominable Pig" (N.Y. Simon & Schuster)
Hart-Davis, Adam	"What the Past Did for Us"
Harting, J. E.	"British Extinct Animals." 40
Hartley-Edwards, Elwyn	"The Country Life Book of Saddlery and Equipment"
Hibbert, Christopher	"Emperors of China"
Hunt, Kathleen	"Evolution of the Horse"
Hyams, Edward	"Animals in the Service of Man"
Hyland, Ann	"Equus" 1990
Hyland, Ann	"The War Horse" 1998
Hyland, Ann	"The Horse in the Ancient World." Sutton Publishing Ltd 2003
Irwin, Robert	"Camel"
Johns, Catherine	"Horses, History, Myth, Art"
Jones Steve	"Almost Like a Whale" 50
Jusserand J.J.	"English Wayfaring Life in the Middle Ages"
Knauer, Elfriede, Regina	"The Camel's Load in Life & Death"
Landels, J.G.	"Engineering in the Ancient World"

Lincoln, Bruce — "Priests, Warriors and Cattle."

Lister, Adrian — "Mammoths" Ice Age Giants

Lorenz, Konrad — "Man Meets Dog"

Lorenz, Konrad — "King Solomon's Ring."

MacDonald, Dr David (ed.) — "The Encyclopaedia of Animals"

Michener, James A. — "Creature of the Kingdom"

Mithen, Steven — "The Prehistory of the Mind"

Moehlman, Patricia D. edited — "Equids: Zebras, Asses and Horses."60

Moorehead, Alan — "Darwin and the Beagle"

Morris, Mary — "Virago Book of Women Travellers"

Morrison, Tony — "Land above the Clouds"

Morrow, William N.Y. 1992 — "The Covenant of the Wild" (Why Animals Chose Us)

Napier, Gordon — "The Rise and Fall of the Knights Templar" 2003, 2007

Ochua, Jorge, Kim MacQuarrie — "Gold of the Andes"

Olsen, Stanley J. — "Fossil Ancestry of the Yak, Its Cultural Significance and Domestication in Tibet"

Olsen, Sandra — "Horses Through Time."

Pickeral, Tamsin — "The Encyclopaedia of Horses and Ponies"

Piggott, Stuart — "The Earliest Wheeled Transport"70

Powell, Keith — "To Hell and Back again for a Camel" (article)

Quinn, Tom — "Shooting's Strangest Days"

Reese, M.M. — "The Royal Office of Master of the Horse"

Renfrew, Colin — "Prehistory"

Ridley, Matt — "Genome"

Rifkin, Jeremy — "Beyond Beef"

Rue III, Leonard Lee — "Wolves" A Portrait of the Animal World"

Russell, Mary — "The Blessings of a Good Thick Skirt."

Shoshani, Jeheskel (Consulting editor) — "Elephants"

Shuker, Karl — "The Beasts that hide from Man"80

Smith, Tilly — "The Real Rudolph"

Steiger, Brad & Sherry	"Horse Mirdacles"
Syroechkovskii, E.E.	"Reindeer"
Tegetmeier, W.B.-M.B.O.U. F.Z.S. & Sutherland, C.L.	"Horses, Asses, Zebras, Mules and Mule Breeding."
Travis, Lorraine	"The Mule"
Trew, Cecil G.	"The Horse through the Ages" 1953
Tomalin, Clare	"Samuel Pepys, the Unequalled Self"
Took, Roger	"Running with Reindeer"
Tyldesley, Joyce	"Myths & Legends of Ancient Egypt"
Van Emden, Richard	"Tommy's Ark"
Walker, Elaine	"Horse" 90
Watson, Lyall	"Elephantoms"
Watson, Lyall	"The Whole Hog"
Whitfield, Susan	"The Silk Road"
Wilson, A.N.	"The Victorians"
Wood, Frances	"The Silk Road"
Wootton, Anthony	"Animal Folklore, Myth and Legend."
(Cambridge:)	"The Development of Transportation in Modern England"
Whitehead, G. Kenneth	"The Ancient White Cattle of Britain & Their Descendants"
Wylie, Dan	"Elephant"
Zeuner, F.E.	"A History of Domesticated Animals" 1963 100
Reader's Digest	"Lost Civilisations"

ABOUT THE AUTHOR

I was born and went to school in South Africa and read English and History at St Andrews University in Fife. As far as contact with the animals in my history I have been in touch with dogs all my life. I have interviewed the head of Canine Partners, the training school for dogs to help the disabled. I have ridden, raced and hunted on horses; have interviewed helpers at 'Riding for the Disabled' in the Cape South Africa. I have ridden camels in Egypt and the Wadi Hadhramaut and attended the Pushkar Camel Fair in Rajasthan. I have also ridden elephants in Nepal and watched herds of them in Africa. I have been in contact with Tilly Smith who runs the Reindeer herd in Aviemore, Scotland. I am also a wildlife sculptor (www. rosiesturgis).

I feel this book would have a wide and timeless appeal not only to animal lovers but also to readers of social history. As I start in prehistoric times and trace the origin of the six chosen animals, it would also be broadening to the minds of our increasingly urbanized young.

I want to write this book because from the moment I thought of it and started researching, I found it more and more fascinating. We get in our cars, catch a train, fly all over the world and few of us think about how we first got ourselves going from hunter-gatherer to highly mobile man. Then there are so many other results... religion, games, war and animals as carers.

Lightning Source UK Ltd.
Milton Keynes UK
UKOW02f2324190915

258910UK00001B/66/P